CONTEXTUAL REALISM

*A Meta-physical
Framework for
Modern Science*

CONTEXTUAL REALISM

A Meta-physical Framework for Modern Science

RICHARD H. SCHLAGEL

PARAGON HOUSE PUBLISHERS

NEW YORK

Published in the United States by
PARAGON HOUSE PUBLISHERS
2 Hammarskjold Plaza
New York, New York 10017

Library of Congress Cataloging-in-Publication Data

Schlagel, Richard H., 1925-
  Contextual Realism.

  Bibliography:
  Includes index.
  1. Science—Philosophy. 2. Knowledge, Theory of.
3. Truth. I. Title.
Q175.S3518 1986   501   86-8146
ISBN: 0-913729-20-5

Dedicated to my wife who inspired this project.

Most of the propositions and questions to be found in philosophical works are not false but nonsensical. Consequently we cannot give any answer to questions of this kind, but can only establish that they are nonsensical....

(They belong to the same class as the question whether the good is more or less identical than the beautiful.)

And it is not surprising that the deepest problems are in fact *not* problems at all.

Ludwig Wittgenstein, *Tractatus,* 4.003

... we should not deprive ourselves of interesting and challenging problems—problems that seem to indicate that our best theories are incorrect and incomplete—by persuading ourselves that the world would be simpler if they were not there.

Karl Popper, *The Self and Its Brain,* p. 62.

Philosophical investigation aims at discovering the nature of knowledge and the properties of reality which make it capable of being known by man.

Enrico Cantore, *Atomic Order,* p. 3.

Contents

Preface

Ever since Galileo, our beliefs about the world have been influenced, in the main, by the results of scientific inquiry. Not only did Galileo's telescopic observations and experimental investigations of terrestrial motions confirm a new vision of the universe and method of inquiry, he also freed the fledgling discipline of "natural philosophy" (or physical science) from the authoritarian control of theology. For even more crucial than the issue of heliocentrism versus geocentrism in his controversy with the Church, was whether natural philosophy was to be the final authority regarding natural knowledge, or whether science was to remain subject to theological control and constraint. Though Galileo was personally humiliated, his defense of the separation of science and theology was soon vindicated.

From that time, science has been not only an autonomous discipline, it has, along with its offspring, technology, been a progressively dominating influence on our lives. This is especially true of the twentieth century. Except for the industrial revolution, the idea of historical progress, and the theory of evolution, man's mode of living and ways of thinking prior to the twentieth century were relatively unchanged from the past; electric power, radios, telephones, automobiles, airplanes, X-ray machines, penicillin, television, nuclear weapons and reactors, computers, lasers, and interplanetary space flights are all "marvels" of the twentieth century. So too are our fundamental theories about the world and our practical techniques for dealing with empirical problems: the special and general theories of relativity, quantum mechanics and particle physics, neurophysiology and pharmacology, molecular biology and genetic engineering, radio astronomy and cosmology, etc. We live in the wake of a scientific revolution in knowledge more pervasive than that which demolished the geocentric, homocentric universe, and we are on the crest of a conquest of nature more extensive than the earlier explorations which led to the discovery of new worlds.

These remarkable, albeit disquieting, developments have had profound effects on man's customary ways of thinking about himself and nature. Often linked to the disclosure of some completely unexpected feature of the universe (such as the constant velocity of light or Planck's universal quantum of action), these discoveries have a dislocating effect on our entrenched assumptions and concepts, necessitating radical revisions of our theoretical frameworks. While scientists have been increasingly involved in the task of clarifying and interpreting the implications of these developments, traditionally this has been a major function of philosophers. In the seventeenth and eighteenth centuries, Locke, Leibniz, Berkeley, Hume, and the incomparable Kant each provided his own analysis of the implications of Newtonian mechanics. Similarly, after Darwin, philosophers as diverse as Nietzsche, Spencer, Bergson, and Dewey were concerned with the philosophical implications of his disturbing new theory of the origin and nature of man.

While prominent contemporary philosophers such as Popper, Nagel, Hempel, Grünbaum, Toulmin, and Feyerabend have also taken scientific developments as their main focus of interest, the much heralded "linguistic turn" in twentieth century Anglo–American philosophy has largely transformed the latter into an exceedingly specialized discipline utilizing either symbolic logic or linguistic analysis in its treatment of philosophical problems. Claiming thereby to have delineated a specific domain of philosophic inquiry, along with a unique methodology—and hence to have attained an independence from the encroachment of the various sciences—they (Quine is an exception) contribute articles of extremely limited scope which are of interest mainly to a small group of analysts restricted to commenting on each other's views. Ironically, while physicists have been proposing "grand unified theories," analytic philosophers have generally eschewed any attempt at a systematic synthesis of knowledge, with the result that their investigations have become increasingly fragmented and isolated from other fields of inquiry—without, it must be added, achieving any greater unanimity of view than previous philosophers. Concerned mainly with analytical precision and logical rigor, their contributions often appear to be sterile and artificial (e.g., "possible worlds semantics"), as well as lacking in breadth of understanding and relevance to other disciplines. Hence, it is not surprising that in contrast with the extensive sway held by such respected philosophers as James, Bergson, Santayana, Whitehead, Dewey, Russell, and Sartre on the intellectual community, contemporary analytical philosophers are practically unknown outside their "narrow world of philosophy," in the words of Gordon Taylor.

The present work represents a return to the older tradition of philosophy in attempting a coherent interpretation of the impact of recent scientific developments on our conventional concepts of knowledge and reality. It is opposed to the current paradigm of analytical philosophy in three respects: (1) it does not focus on language as the primary datum for philosophical investigation, but the results of empirical inquiry (would scientists have made much progress if they had attended only to the mathematical formalism, ignoring the findings of experimental investigation?); (2) it does not rely on logic as the tool for clarifying concepts and arguments; and (3) it is systematic, in that it attempts to show that the various facets of the problem of knowledge—such as the levels of empirical inquiry, the paradoxical consequences of quantum mechanics, the investigations in neurophysiology, the mind–body impasse, the problems of linguistic reference, and the meaning and criteria of truth—are interrelated and imply a new framework of interpretation: "contextual realism."

Though addressed to my colleagues in philosophy, it is hoped that this book will reach a wider audience comprised of scholars of diverse backgrounds intrigued by fundamental questions transcending their individual disciplines, and even the general reader who would like to acquire a better understanding of how current research in the sciences affect our conceptions of knowledge and reality. Unlike what he would find in many books dealing with similar topics, the reader will not here encounter the logical symbolism and technical argument that constitute a permanent barrier to understanding. The text is written as clearly as possible, in a style that will discourage no one from reading because of specialized notation or exclusivity of subject matter. Whatever its remaining defects, the text has been much improved thanks to the revisions suggested by the editors of Paragon House, to whom the author is especially grateful.

While addressing current epistemological and linguistic issues, as well as the views of the main contributors to these issues, no attempt is made to provide a detailed analysis of all the related positions. Believing that analytical discussions all too often pertain to extremely marginal counter–examples, so hypothetical and remote from actual experience as to be of dubious significance, this book concentrates on the central problems themselves, along with the pertinent empirical evidence. In the opinion of the author, one fact is worth a thousand philosophical arguments, however ingeniously contrived. If philosophy is to regain something of its traditional prominence, it must engage problems closer to their empirical source rather than just ascend to the rarefied atmosphere of linguistic analysis.

Richard H. Schlagel
The George Washington University

Introduction

No problem has been more central or crucial to the history of philosophy than the problem of knowledge. From the time of the ancient Greeks, from whom we derive the term 'epistemology' (theory of knowledge), questions regarding the nature, basis, and limits of knowledge have been at the forefront of philosophical inquiry. Essentially, the problem arises because the world is not self-explanatory—if it were, there would be no problem of knowledge. Every entity and event would display the reason for its being or becoming, so that mere awareness of the world would constitute knowledge—although one would hardly be conscious of possessing knowledge for there would be nothing to contrast it with. Such a world would be very different from ours.

If knowledge was mothered by curiosity, as Aristotle claimed, then surely it was fathered by ignorance. How fragmentary, limited, and superficial our experience and knowledge is when contrasted with nature's seemingly unbounded extent and complexity. In spite of the tremendous progress made recently in the physical sciences, a fundamental understanding of even the most familiar phenomena still eludes us. For example, it is common knowledge that an unsupported object falls to the ground. But what causes this fall? The familiar answer is gravity; but what is gravity and how does it operate instantaneously at a distance? Is it an inexplicable force, as Newton believed, or is it identical with the structure of space–time, as Einstein proposed? Similarly, how is one to explain our perception of color? Are colors, as they appear to be, independently existing surface qualities of physical objects, or is their occurrence dependent upon complex radiational and neurological processes to that they could no more exist independently of the organism that perceives them than sensations of heat or pain? Or, should one wish to maintain both theses, how are they to be reconciled?

Just as nature does not immediately disclose its *raison d'etre,* so the sensory qualities, physical properties, and diverse processes that it manifests occur as if they were self–contained and self–sufficient. The sun emits light and heat, the planets revolve, thunder claps and lightning strikes, plants grow, fire burns, children learn languages, and people think as if there were no deeper elements, structures, or causal factors involved. Everything occurs as if nature were completely displayed in its surface appearances. Despite the extensive processes necessary to support the biological existence of a human being (as well as his motor activities and conscious experiences), we are normally unaware of their occurrence. Today, although we realize that the properties of objects and the actualizing of events depend upon complex inner structures and sustaining background conditions, we are oblivious of this as long as these conditions and structures persist, camouflaged as they are by the surface manifestations and processes of nature. It is due to this foreshortened appearance of nature that the need for experimental inquiry to uncover these deeper domains was relatively late in being recognized. For similar reasons, Plato and Aristotle, although they were aware that natural appearances and occurrences were in need of further explanation, sought this explanation in the manifest forms and qualities of nature itself, however abstract and idealized. As a result, their theories of knowledge were realistic and essentialistic.

According to Plato, for example, though the empirical world of the Receptacle was subject to continuous change and dissolution, and thus was too imperfect to be an object of knowledge, still, it was an imperfect representation of the unchanging, perfect, eternal Realm of Forms. These Forms served as archetypes for creation by the Demiurge, as well as the referents of universal, mathematical, and value terms (such as 'whiteness,' 'unity,' and 'goodness,'), providing the objective meanings for a common discourse—the origin of the "name theory of meaning." In addition, since for Plato one could only have knowledge as such of the Real, the Eternal Forms as the objects of an intellectual intuition guaranteed truth, while our judgments about empirical events were merely belief or opinion. Knowing was analogous to seeing: a direct vision of the independently real Forms. Thus, while the ordinary empirical world of change and becoming was merely an imperfect reflection of the Realm of Forms, on the other hand, the Realm of Forms was an eternal prototype or exemplar of the things found in the empirical world, while including such unique entities as mathematical objects and the Form of Goodness.[1]

Consistent with this view of knowledge, Plato rejected the conception of applied mathematics as it was being developed by the Pythagoreans, instead conceiving of mathematics as an intuitive (or mystical) disclosure of the ultimate harmony of the universe. This vision is vividly expressed in the *Epinomis,* a work which, if not written by Plato himself, is directly in the Platonic tradition.

> To the man who pursues his studies in the proper way, all geometric constructions, all systems of numbers, all duly constituted melodic progressions, the single ordered scheme of all celestial revolutions, should disclose themselves, and disclose themselves they will, if, as I say, a man pursues his studies aright with his mind's eye fixed on their single end. As such a man reflects, he will receive the revelation of a single bond of natural interconnection between all these problems.[2]

The closest Plato came to a modern explanation of the changes of phenomena (as opposed to a unified mathematical formalism) was in the *Timaeus* where he reduced the standard four elements, air, earth, fire, and water, to the four Pythagorean solids, explaining their transactions as due to the breakdown and exchange of the three kinds of triangles making up the four geometric solids.[3] Here is a visionary anticipation, however *a priori* and conjectural, of the modern explanation of chemical reactions and of the steriometric structure of matter.

Compared with Plato, Aristotle's theory of knowledge was more empirical in that he did not separate the Forms from the empirical world, and maintained that all knowledge was initially dependent upon sensory induction, though "scientific knowledge" as such consisted of deductive demonstration. But even more than Plato, Aristotle believed that the world could be understood in terms of directly manifested qualities, forms, and processes. In fact, the observable forms of things constituted their essential natures along with determining their specific actualizations and functions. As he maintained, the stars move in a continuous, eternal, circular motion because they are composed of a fifth substance, the incorruptible, weightless, celestial ether, while the four terrestrial elements, fire, earth, air, and water, have a natural rectilinear motion, with a beginning and an end, characteristic of terrestrial phenomena. Earth and water have an inherent tendency, because of the kinds of elements they are, to fall to their natural place in the center of the universe, while air and fire naturally rise to the circumference of the sublunar world. In this model, there is a necessary connection between the qualities or forms of things and their behavior.

Moreover, Aristotle believed that all the diverse phenomena of the universe could be classified into natural kinds: species and genera. Each entity had its specific characteristics and behaved the way it did because of the type of entity it was. Explaining a phenomenon consisted of demonstrating that it had a particular trait because it belonged to a class of entities which necessarily possessed that trait. Socrates, for example, was mortal because he belonged to the genera of animals which are by nature mortal, while he behaved rationally because he belonged to the species man which is inherently rational. Similarly, planets do not twinkle (as stars do) because they belong to the class of proximate (stellar) bodies which by nature do not twinkle. In fact, Aristotle maintained that the *cause* of phenomena was equivalent to the *connecting middle term* of the premises of a scientific demonstration, in which the questionable attribute is demonstrated as necessarily belonging to a particular entity because of that entity's inclusion in a higher genus. Change and becoming were explained as due to the transition from potentiality to actuality when a thing actualizes its essential nature under the appropriate conditions. An acorn develops into an oak tree, given the proper conditions, because it possesses the form of oak tree potentially within it. All change is regulated by the fact that things can only lose and take on certain properties, privative and positive contraries, because of the kinds of things they are: a ripening olive, for example, changes from green to black, while a tomato changes from green to red because of the kinds of fruit they are.

From the contemporary standpoint, Aristotle's explanations seem essentially classificatory, especially as he rejects the modern scientific mode of explaining phenomena in terms of underlying, unobservable substructures. Although in his biological treatise he advocated dissection to enhance one's knowledge of organisms, generally he relied uncritically on ordinary observation as the basis of his explanations. Thus, he claimed erroneously that falling objects accelerate in proportion to their weights, and that a terrestrial object, when put in motion, naturally comes to rest, from which he deduced that a continuous mover is necessary to account for continued motion (in contrast to the principle of inertia, and Newton's law that *additional* force results in *additional* increments of motion or acceleration). Nevertheless, Aristotle's brilliant empirical

observations and reflections were first approximations of a more
exact knowledge, explaining why Galileo begins his study of motions with a critique of Aristotle's analysis. Though today Aristotle's
overall scientific framework resembles a common sense description
based on the superficial, observable processes of nature, that he
could achieve such an objective, empirical outlook at that time in
history is a remarkable achievement.

In contrast to Plato and Aristotle, the ancient atomists, Leucippus
and Democritus (following Empedocles and Anaxagoras), developed a model of explanation characteristic of nearly all modern
science, attempting to explain the manifest qualities, processes, and
changes in the world in terms of the motions of underlying, imperceptible elements, the atoms. However crude in detail, the atomists
system presaged the fundamental categorical scheme of the modern
mechanistic word view. Purging the universe of all mythical, tele-
ological, and qualitative aspects, their general explanation of phenomena was in essential respects identical to that of Newtonian
science. In place of the ordinary macroscopic objects of experience,
the real elements of the world were conceived as imperceptible,
indivisible atoms defined in terms of such tactual-visual properties
as shape, size, solidity, and motion. As in Newtonian mechanics, all
observable displacements and changes were attributed to the *underlying* motion and impact of these basic elements forever moving
in an independent, endless space. Anticipating both the Kant–
Laplace nebular hypothesis and the contemporary steady–state
theory of the universe, the Atomists explained the origin of particular worlds as due to the chance interaction and consequent
aggregation of individual atoms, while the universe as a whole
consisted of innumerable worlds in various stages of formation
and disintegration.

Whereas the philosophies of Plato and Aristotle provided a guarantee of objective knowledge but a weak explanation of natural
phenomena, the Atomists provided a more adequate general framework (as attested to by later developments in science) for explaining natural phenomena, but raised intractable questions for a
general theory of knowledge that have challenged modern philosophy. The following well-known fragment, attributed to the Atomists
by Sextus, states the problem very clearly:

> By convention are sweet and bitter, hot and cold, by convention is
> colour; in truth are atoms and the void....In reality we apprehend
> nothing exactly, but only as it changes according to the condition of
> our body and of the things that impinge on or offer resistance to it.[4]

Slightly more than twenty centuries later, the same thesis was asserted by Galileo in practically identical words:

> ... I think that these tastes, odours, colours, etc. on the side of the object in which they seem to exist, are nothing else than mere names, but hold their residence solely in the sensitive body; so that if the animal were removed, every such quality would be abolished and annihilated.[5]

According to the view of experience expressed in these quotations, we do not perceive the world as it truly is. Recognizing that perception occurs as a result of the stimulation of our external senses, and believing that the various physical stimuli are in some way transformed by the nervous system and/or the mind into experienced sensory qualities, scientists and philosophers drew a distinction between the independently existing physical properties of things and the qualitative effects they produce in the organism. In spite of the *perceptual* identity of colors, smells, tactual sensations, heat and the like with actual things (in the sense that we do not experience such qualities as occurring in our heads or in our minds, but as qualifying real things in the outer world), a *conceptual* distinction was made between the sensory qualities an object appears to have when interacting with an organism such as a human being, and the physical properties it possesses independently of any such interaction. That an object should appear different or exhibit unexpected properties under different conditions or within various contexts is one of the essential features of the modern conception of the physical world, in contrast to the realistic and essentialistic conceptions of Plato and Aristotle.

Having made the distinction between the object as it is in itself and as it appears to us, the problem became, (1) to define the independently existing object, (2) to describe the effects it has on the organism, (3) to explain how these physical and neurological effects are experienced as sensory qualities and, (4) to justify a knowledge of the world not merely as it appears to human beings but as "it is in itself", independent of them.

Attempts to solve the first problem have been pursued by physicists and chemists, initially in terms of indivisible atoms or particles defined by means of primary qualities such as those indicated by the early Atomists; then, when the conception of an indivisible atom proved inadequate at the turn of the century, in terms of electrically charged particles modeled after our solar system. Although the atom has resisted final analysis or description, the latter model has proven eminently successful as the basis of theoretical and experimental research in atomic physics, chemistry, astronomy, molecular biology, and related fields.

The second problem—how do physical objects transmit their stimuli and affect the various sense organs, and how are these affects transmitted through the nervous system to the brain—has been relegated, since the time of Descartes and Locke, to the neural sciences for investigation. Indeed, tremendous advances have been made in neurophysiology during the present century describing neurological processes in terms of fundamental chemical—electrical transmissions.

In comparison with these problems, the last two mentioned above were taken over principally by philosophers and, despite much ingenuity expended, have remained intransigent. The question as to how chemical–electrical neuronal discharges eventuate in conscious experiences or, more generally, how the mind is related to the nervous system and brain of the organism is, of course, the baffling mind–body problem. The ancient Atomists, and later Hobbes, attempted to construct a unified theory according to which conscious processes were nothing more than another type of atomic process (though the atoms involved were smoother and more mobile). More recently, advocates of extreme reductionism such as J. J. C. Smart and D. M. Armstrong have also denied conscious processes as such, in support of a completely physicalistic account of the universe, maintaining that what we call conscious experiences are simply neurological processes and *nothing more.*[6]

This extreme monism has been adopted in desperation to avoid a conception of reality that would include two irreconcilable and irreducible components, the mental and the physical. Although this dualism is of very ancient origin, its modern formulation is due, in particular, to Descartes. As a founder of the modern mechanistic view of the world, Descartes believed that the physical world could be reduced to one essential property, extension. This position required that the other qualities seemingly possessed by objects, such as hardness; color, odor, texture, and heat, be given another location 'in' the mind or the soul. Unlike the reductionists, Descartes did not believe that conscious thought (including such innate, clear and distinct ideas as "cogito ergo sum" and the conception of God) could be explained in mechanistic terms, hence he attributed it to another substance, mind, soul, or spirit. In addition, Descartes offered the ingenious *"cogito"* argument showing how the entire apparent world—in contrast to the real one of extension—could exist within the mind.

Since Descartes' dualism has been much maligned by ordinary language philosophers such as Gilbert Ryle, who argued that it consisted of a simple "category mistake,"[7] it is worth treating Descartes with some care, especially since his argument redirected the whole of modern philosophy. Although his major influences were mathematical physics and Augustian Neoplatonism, as a contemporary of Galileo one can conjecture that Descartes was not entirely unaware of the controversy engulfing Galileo's support of the Copernican world view. One of the subtle aspects of this fascinating but tragic controversy was the question of God's veracity. For if Copernicus' heliocentric theory were correct, then the rising and setting of the sun, as well as the centrality and motionlessness of the earth, would be deceptive or illusory. This implied that our senses deceive us, and if we were created by a perfect being, that being must be responsible for this deception.[8]

Galileo's telescopic observations, which provided the first new astronomical data since antiquity, reinforced this conclusion by showing that our senses were limited as well as deceptive. Would God have deliberately created man with imperfect and deceptive senses? While for reasons of prudence Descartes never refers to the Copernican controversy, instead invoking dreams, sensory illusions, and errors of perceptual judgment as grounds for distrusting the senses, his references to a "deceitful demon" could apply to the current astronomical controversy as well.

> If...it is contrary to His Goodness to have made me such that I constantly deceive myself, it would also appear to be contrary to His goodness to permit me to be sometimes decived, and nevertheless I cannot doubt that He does permit this....I shall then suppose, not that God who is supremely good and the fountain of truth, but some evil genius not less powerful than deceitful, has employed his whole energies in deceiving me....[9]

If our perceptions could deceive us about something as obvious as the motionlessness of the earth and the apparent diurnal rotation of the heavens, where should we turn for a more secure foundation of knowledge? For Descartes, the alternative was clear. Knowing our senses deceive us in some instances, we have no assurance they do not deceive us in all, therefore we must seek a more reliable basis of knowledge in self-evident principles analogous to those of logic and mathematics—the kind of principles and reasoning used by Copernicus in setting forth his astronomical theory, and by Descartes himself, with extraordinary success, in creating analytical geometry. Ordinary language philosophers such as Gilbert Ryle, John Austin, and G. J. Warnock have criticized the inference that because our senses deceive us some of the time they might deceive us all of the time, on the grounds that it violates the ordinary distinction between veridical and non-veridical perceptual judgments. Arguing the converse of Descartes, they claim that if our senses deceived us all of the time we would have no basis for saying that they deceived us some of the time, because the latter claim presupposes a distinction between deceptive and reliable perceptions. This criticism has some validity, but it fails to take into account the fact that while one can distinguish between veridical and non-veridical perceptions *within* the context of ordinary experience, the possibility that ordinary experience itself might mis-prepresent the world is not precluded. For example, while pilots make various navigational computations and judgments within the context of the geocentric view (which is assumed to be valid for most terrestrial navigational purposes), this does not prove the ultimate truth of geocentrism.

The example of dreams is especially important in exhibiting the thrust of Descartes' argument. As he clearly argued, the essential point about dreams is, if one assumes that our ordinary perceptions disclose the real world, then it is difficult to understand how one can have dreams which for many people while dreaming (and even for some time afterwards), appear indistinguishable from waking experience. That is, if as we normally assume, a real world must be present to be experienced, how can we have experiences indistinguishable (at the time) from normal experiences, when the contents or referents of those experiences are not present? If one is not really being chased by a madman with a knife, how can one have such a terrifying, life-like experience of being chased (perspiration, shortness of breath, increased heart beat, and aborted attempts to cry out)? How does the content of the dream acquire that vivid, compelling reality *within the context of the dream*? Like it or not, Descartes posed a problem which cannot be dismissed merely by pointing to a common linguistic distinction. In the absence of the usual physical causes, what *does* enable us to have experiences which, though not as complete, are still qualitatively similar to normal ones? If one resorts to the usual contemporary explanation that the (unconscious) cortical activity of our brains causes us to have these experiences, then it is at least possible this same activity could be primarily responsible for the qualitative content of our waking experience as well, though normally dependent upon external stimuli for its activation.

Having pointed up the importance of sense illusions, errors of perceptual judgment, and the qualitative similarity between the contents of dreams and those of waking experience, Descartes asked us to consider whether, instead of having been created by an omnipotent, omniscient, all-perfect God, we might not have been created by a malevolent demon who had deliberately set out to deceive us—and if so, was there any way of extricating ourselves from this deception? Unfortunately, Descartes' solution, depending upon the infallibility of self-evident principles, proved deficient, because if we can assume a demon created us with deceptive senses, we can also suppose he created us susceptible to being deceived (as we sometimes have been) by seemingly self-evident principles and rational demonstrations, (the "*cogito ergo sum*" argument notwithstanding). For just as we could dream of someone raising the "*cogito*" argument within a dream (and some people claim they have), and nevertheless not conclude that such a person truly exists, so we could be the figment of a demon's imagination and not really exist ourselves (at least not as independent thinking substances, as Descartes thought he had proved).

But Descartes' conclusions aside, what is important about his argument, and what established him as the father of modern philosophy, is the fact that he introduced a new "paradigm"[10] of philosophical inquiry, a novel cognitive frame from which the problems of knowledge were seen in new perspective. For of primary significance regarding the "*cognito*" argument is Descartes' inference that one could know one's own consciousness and its contents with greater certainty than one could know anything else: that *knowledge* of one's own consciousness must be *epistemologically prior* to knowledge of anything else, and hence the *existence* of the mind must be *ontologically independent* of anything else.

That is, even if our sensory experiences are deceptive in the sense that they do not conform to an independently real world (as is true of dreams and certain drug-induced hallucinations), one cannot deny having the experiences themselves and therefore such sensory contents must exist, at least within the mind. Moreover, since this mode of analysis raises a doubt as to whether there is a real world at all, while maintaining the certainty of our knowledge of the mind's existence, the latter cannot depend upon the former. (This is an excellent example of the kinds of philosophical arguments that underlie philosophical positions.) As Descartes states:

> But it is very certain that the knowledge of my existence taken in its precise significance does not depend on things whose existence is not yet known to me....I am the same who feels, that is to say, who perceives certain things, as by the organs of sense, since in truth I see light, I hear noise, I feel heat. But it will be said that these phenomena are false and that I am dreaming. Let it be so; still it is at least quite certain that it seems to me that I see light, that I hear noise and that I feel heat. That cannot be false....[11]

Though the conviction that the contents of perception are mind-depend, and thus ontologically independent of the external world, seems opposed to everything in our everyday experience, this paradigm redirected philosophical inquiry from the time of Descartes to the beginning of the twentieth century when it came under attack in the classic articles of G. E. Moore[12] and Ralph Barton Perry.[13] From the time of Descartes, the main philosophical traditions took for granted that one had to begin all philosophical investigation with the immediate contents of consciousness ("ideas," "sense impressions," "the sensory manifold of intuition," or "sense data"), described by Richard Rorty as "the veil of ideas" or "the glassy essence," and thus that all knowledge of an independent world had to be inferred from these mind-dependent, conscious contents. Bertrand Russell codified this assumption in his extremely influential distinction between "knowledge by acquaintance" and "knowledge by description,"[14] while Whitehead baptized it "the bifurcation of the world" into mind and matter. The latter was an appropriate designation since Descartes had defined matter as an extended, causal, publicly observable substance, while mind was depicted as an unextended, non-causal, privately known spiritual substance. This dualism was another aspect of the new paradigm introduced by Descartes.

In contrast to the theories of Plato and Aristotle, Descartes' dualistic, subjectivistic conception of our experience in relation to the real world raised directly the fourth and last problem referred to earlier. If all we directly experience are the effects of an independent world on our sense organs communicated by mechanical processes to the brain where they are transformed and experienced by the mind as sensations of heat, roughness, hardness, and so forth, as well as sense qualities like colors, sounds, and smells, then how can we transcend this sensory barrier to arrive at a knowledge of the independently real world? Though all the leading scientists of the seventeenth and eighteenth centuries were in the same "predicament," because they also accepted the subjectivity of sensory data, they refused to be deterred, proceeding on the assumption that there must be a real world and that we are capable of knowing something about it. Newton, for example, although conceiving of the physical world as composed of matter formed as imperceptible "solid, massy, hard, impenetrable, movable particles," surmounted the problem in his third rule: "The qualities of bodies, which admit neither intensification nor remission of degrees, and which are found to belong to all bodies within the reach of our experiments, are to be esteemed the universal qualities of all bodies whatsoever."[15] That is, those qualities which are universally found in objects and which cannot be separated from them under any conditions (either physically or conceptually) are to be considered the real qualities of objects, even of those objects imperceptible to the senses. Boyle argued in a similar way that

> since experience shows us that...matter is frequently made into insensible corpuscles or particles, we may conclude, that the minutest fragments, as well as the biggest masses of the universal matter are likewise endowed, each with its own peculiar bulk and shape....[16]

While scientists such as Galileo, Newton, and Boyle refused to be deterred by the epistemological problems inherent in the causal explanation of perception, psycho–physical dualism, and the discontinuity between the ordinary macroscopic world and the atomic conception of physical reality, the philosophers took these problems to heart. As a result, they constituted the primary data of philosophical inquiry, at least until the middle of the present century. There is no need to recount in detail the various attempts to resolve these problems: how Locke, starting with "ideas" as "the sole objects of consciousness," nevertheless maintained (with the scientists) that a valid distinction could be drawn between primary and secondary qualities (the former truly inhering in the independently existing insensible particles of the world, and the latter constituting the capacity or "powers" of the primary qualities to affect other things) and sensations and ideas in our minds; how Berkeley maintained that an idea could only be like another idea, hence *"esse est percipi,"* or to be either is to be perceived or to be the mental or spiritual cause of what is perceived; how Hume drew out the skeptical conclusions, asserting that while on the *practical* level it would be absurd to deny causality and the reality of an independent world, on the *theoretical* level, both beliefs in causality and in a real world were merely conditioned psychological states (Santayana called such beliefs "animal faith") having no theoretical justification; how Kant denied the possibility of any knowledge of "things in themselves" transcending the phenomenal world, claiming that we can have scientific knowledge of things only as they appear to us due to our cognitive makeup; how Mach tried to reduce all scientific knowledge to an economical description and prediction of phenomena, maintaining that ordinary objects were simply a construction from sense elements, and that appeal to an imperceptible physical domain as cause of phenomena was merely a "convenient fiction" to facilitate prediction; and finally, how John Dewey denounced all dualisms, including an "antecendent reality" lurking behind ordinary objects, maintaining the instrumentalist position that only the predictive consequences of scientific hypotheses in ordinary experience were of real concern, not their ultimate truth or ontological status.

As these last two positions indicate, much of twentieth century philosophy has been characterized by the belief that all *genuine* problems are solvable, either by the present techniques of science or by their continued development, and hence that the traditional problems of philosophy, subject to endless dispute, are "metaphysical" or "pseudo-problems." Using as example the outstanding success of *Principia Mathematica,* which set forth the foundations of mathematics (over which there is much disagreement today, particularly concerning "The Axiom of Reducibility") with the newly developed techniques of symbolic or mathematical logic developed by Frege and himself, Bertrand Russell conveyed to philosphers the confident belief (although he did not entirely succumb to it himself) that philosophical problems could be similarly resolved. Moreover, as Russell thought he had illustrated in his influential article, "Theory of Descriptions,"[17] metaphysical questions, such as those having to do with the putative reality of abstract entities, often arise from confusing the superificial *grammatical form* of propositions with their deeper *logical form*. As Wittgenstein stated in the *Tractatus* (4.0031): "It was Russell who performed the service of showing that the apparent logical form of a proposition need not be its real one." Misled by the grammatical structure of language, philosophers, according to Russell, often posed problems and developed arguments which, when considered in relation to the actual *logical form* of propositions, can be seen to be artificial or misleading. Again, as Wittgenstein asserted in the "Preface" to the *Tractatus,* "the reason why these problems are posed is that the logic of our language has been misunderstood."

To the thesis that philosophical problems are not substantive issues but have their origin in linguistic ambiguities and confusions, the logical empiricists or positivists added the claim that most philosophical propositions are literally "nonsensical" because they are not amenable to sensory verification or falsification. In contrast to scientific propositions, philosophic assertions are not empirically testable and are therefore meaningless. Carnap proposed that an ideal language be developed meeting the requirements of empirical significance, but also conforming to specified linguistic and logical rules, with the purpose of eliminating possible linguistic ambiguities and purported metaphysical statements.[18] In this way it was hoped one could avoid the type of problems and irresolvable disputes that had plagued traditional philosophy. But even if an ideal language serving these purposes could have been developed (which did not prove to be the case), this would not explain the fascination such problems have held for philosophers in the past. Moreover, the thesis that philosophers (such as Hegel, Bradley, and Whitehead) had been spouting nonsense all their lives without realizing it, hardly seemed credible.

Accordingly, it fell to another school of philosophy, ordinary language philosophy, by far the most prominent group among Anglo–American philosophers since the Second World War, to provide a more plausible analysis. Having been influenced by the lectures and writings of G. E. Moore and particuarly by the post–*Tractatus* seminars and privately circulated notebooks of Wittgenstein (which emphasized the crucial role of ordinary language in the formulation of philosophical arguments), ordinary language philosophers claimed that philosophical problems do not arise from a misinterpretation of the underlying *logical form* of propositions, but from misusing *ordinary language* itself, the language in which philosophy is, for the most part, expressed. Misconstruing the 'logic' inherent in ordinary linguistic uses, philosophers raise unnecessary philosophical puzzles. According to the revised view of philosophy presented in Wittgenstein's *Philosophical Investigations*: "Our investigation is therefore a grammatical one. Such an investigation sheds light on our problem by clearing misunderstandings away. Misunderstandings concerning the use of words. . . ."[19]

On this view, philosophical inquiries do not actually arise from reflecting on such problems as the causal conditions of perception, the status of macroscopic objects versus the theoretical entities of science, or the relation between conscious experiences and their underlying neurological processes, as believed by past philosophers, but are caused, instead, by linguistic misunderstandings. It is not the human condition and the various anomalies in knowledge that are the source of philosophical problems, but linguistic muddles caused by the misuse of ordinary language. Again, to quote from the later Wittgenstein:

> The problems arising through a misinterpretation of our forms of language have the character of *depth*. They are deep disquietudes; their roots are as deep in us as the forms of our language and their significance is as great as the importance of language.—Let us ask ourselves: why do we feel a grammatical joke to be *deep*? (And that is what the depth of philosophy is.)[20]

Philosophical problems are similar to grammatical jokes! They are as deep as language, but not as deep or as puzzling as the universe. Linguistic analysis has been compared to psychoanalysis, both having as their function the elimination of disturbing problems by tracing them to unsuspected origins, showing, thereby, that the problems do not have the significance they appear to have for the individual on the immediate conscious level.[21]

Whereas scientists in the twentieth century have found themselves at times overwhelmed by the seemingly unfathomable "paradoxes" and insurmountable "impasses" in their investigations, ordinary language philosophers (along with the positivists) conveyed the impression that there really is nothing very perplexing or mysterious about our existence in and knowledge of the world. For instance, in contrast to Werner Heisenberg who asked in desperation, "can the world possibly be as absurd as it appears to us in our scientific theories?"[22] and to Eddington who pondered the vast differences between the ordinary macroscopic world and the scientific picture of physical reality,[23] G. E. Moore states, "I do not think that the world or the sciences would ever have suggested to me any philosophical problems."[24] Ordinary language analysts were fond of pointing out that, in contrast to philosophers, "the plain man" does not find himself beset by the kinds of problems philosophers find perplexing—as if there were no difference between the two! Moore also claimed that he had solved the problem of the reality of physical objects by pointing to his hand. In a similar manner, Ryle thought he had disposed of the mind–body problem by tracing it to a "category mistake," and proposed to dispense with the question of the status of scientific entities, vis- à–vis ordinary physical objects, by maintaining that they were not intended to be duplicate (or independently real) entities, but simply numerical notations similar to those on library cards indexing the locations of books.[25] But how one could be persuaded that chemical molecules, hydrogen atoms, atomic radiation, X–rays, and the like are not independently real, but merely numerical notations of ordinary objects, is almost beyond comprehension.

For those who are convinced that the only genuine problems are those pursued by the sciences—in contrast to speculative questions engendered by fundamental social, cultural, and intellectual changes in society, such as political, scientific, and technological revolutions—and therefore, that philosophical inquiry should be restricted to an analysis and description of ordinary linguistic uses (involving "family resemblances" or "language games"), somewhat misleadingly referred to as "conceptual analysis," there is little point in proceeding. This book is written for a new generation of philosophers that is, it is hoped, beginning to find the older paradigm of analytic philosophy inadequate to deal with the complex theoretical problems generated by empirical inquiries, technological innovations, social changes, and intellectual revolutions.

Although the brief history of philosophy sketched in this Introduction should be well known, the purpose in recounting it was to remind the reader that modern problems of philosophy had their origin in scientific developments of the seventeenth and eighteenth centuries, not in an inadvertent misuse of ordinary language. No adequate understanding of the founders of modern philosophy from Descartes to Kant is possible without relating their thought to the new science of mechanics. Contemporary philosophers face a similar challenge in attempting to analyze, clarify, and assimilate the consequences of the profound social, intellectual, and technological upheavals that have occurred in the twentieth century.

Accordingly, this book will take seriously the traditional problems of philosophy, such as the problem of perception, the mind–body problem, and the question as to the nature of the real world, in attempting to develop a theory of knowledge and a conception of reality consistent with the remarkable developments of twentieth century science. The advantage that a contemporary philosopher has over his seventeenth and eighteenth century predecessors is that recent scientific discoveries have provided him with new data and insights for dealing with the traditional problems. Five main scientific advances which lay beyond the horizon of classical modern philosophers indirectly affected by their absence, the ways in which these philosophers approached their problems: (1) the theory of evolution, (2) the revolutionary developments in mathematics (such as the discovery of non–Euclidian geometries) and physics, particularly relativity theory and quantum mechanics, (3) the development of experimental psychology with its empirical methods for investigating cognitive structures, (4) the striking progress made in this century in neurophysiology, and (5) the fascinating developments in linguistics and in artificial intelligence.

Because none of the modern classical philosophers could think of man as an organism more highly evolved than other creatures, or could conceive of nature as anything but a vast deterministic machine, or could use any other method than introspection when considering the origins and foundation of knowledge, or could anticipate the progress made in correlating conscious processes with intricate patterns of nerve firings, or was aware of the crucial role of language in philosophical inquiry, his treatment of problems was limited. Taking account of these new developments could significantly modify the traditional paradigms within which the problems of knowledge were formulated and the various strategies conceived for dealing with them. In fact, we might well be reaching the end of the Cartesian era.[26] This is not to imply, however, that because some of the previous limits of knowledge have been displaced we can expect ultimate solutions to these problems, or that our present conjectures will not be similarly restricted by a lack of knowledge which lies beyond existing frontiers of inquiry. On the contrary, it will be an essential thesis of this book that all experience and knowledge are relative to various contexts, whether physical, historical, cultural, or linguistic, and that as the contexts change, so do the perspectives one has on these problems. What keeps man from the plight of Sisyphus is that though the (intellectual) burden remains essentially the same, at least the context changes, so mankind is not condemned to the same fate of repetition.

Notes to introduction

1. This discussion of ancient Greek philosophy is based on my book, *From Myth to the Modern Mind: A Study of the Origins and Growth of Scientific Thought,* Vol. I, *Animism to Archimedes* (Bern: Peter Lang, 1985).
2. *Epinomis,* 991e–992a, trans. by A. E. Taylor.
3. Cf. Plato, *Timaeus,* 48b–56d.
4. Sextus, *Advanced Mathematics,* VII, 135. Quoted from G. S. Kirk and J. E. Raven, *The Presocratic Philosophers* (Cambridge: Cambridge University Press, 1957), pp. 422, n. 589.
5. Galileo Galilei, *Opere,* IV, 333ff. Quoted from E. A. Burtt, *The Metaphysical Foundations of Modern Science,* rev. ed. (Garden City: Doubleday & Co., 1954), p. 85.
6. Cf. C. V. Borst (ed.), *The Mind–Brain Identity Theory* (New York: St. Martin's Press, 1970).
7. Cf. Gilbert Ryle, *The Concept of Mind* (London: Hutchinson's University Library, 1949), ch. I.
8. Cf. Ludovico Geymonat, *Galileo Galilei,* trans. by Stillman Drake (New York: McGraw–Hill, 1965), p. 51.
9. René Descartes, "Meditations I," trans. by E. S. Haldane and G. R. T. Ross. Compare this discussion with that of Richard Rorty in *Philosophy and the Mirror of Nature* (Princeton: Princeton University Press, 1979), ch. 1.
10. Cf. Thomas S. Kuhn, *The Structure of Scientific Revolutions,* 2nd ed. (Chicago: University of Chicago Press, 1970). While Kuhn made the concept of "paradigm" well known in his influential analysis of scientific developments, the term can be used to designate the "programs" of various schools of philosophy as well.
11. René Descartes, "Meditation II," trans. by E. S. Haldane and G. R. T. Ross.
12. Cf. G. E. Moore, "The Refutations of Idealism," in *Philosophical Studies* (London: Routledge & Kegan Paul, [1922] 1951), pp. 1–30. Originally published in *Mind,* N.S. vol. xii, 1903.
13. Cf. Ralph Barton Perry, "The Ego-Centric Predicament," in *Present Philosophical Tendencies* (New York: Longmans, Green & Co., 1929), p. 129ff.
14. Cf. Bertrand Russell, *The Problems of Philosophy* (London: Oxford University Press, [1912] 1952), ch. V.

15. Isaac Newton, "Rules of Reasoning in Philosophy," *Philosohiae Naturalis Principia Mathematica,* 1686, Bk. III, rule III.
16. The Works of the Honourable Robert Boyle, Birch (ed.), 6 Vols., London, 1672, Vol. III, p. 16. Quoted from E. A. Burtt, *Metaphysical Foundations of Modern Science, op. cit.,* p. 174, N. 38.
17. Cf. Bertrand Russell, "Theory of Descriptions," *Introduction to Mathematical Philosophy* (London: George Allen & Unwin, 1919), ch. XVI. F. P. Ramsey called Russell's theory a "paradigm of philosophy."
18. Cf. Rudolph Carnap, *Der Logische Aufbau der Welt* (Berlin: Weltkrus–Verlag, 1928), and *Logische Syntax der Sprache* (Vienna: Springer, 1934).
19. Ludwig Wittgenstein, *Philosophical Investigations,* trans. by G. E. M. Anscombe, 2nd ed. (Oxford: Basil Blackwell, 1958), Part I, sec. 90.
20. *Ibid.,* Part I, sec. 111.
21. Cf. Morris Lazerowitz, *The Structure of Metaphysics* (London: Routledge & Kegan Paul, 1955).
22. Werner Heisenberg, *Philosophy and Physics,* Vol. XIX of World Perspective Series, 1958. Quoted from the Harpers Torchbook ed., 1962, p. 42.
23. Arthur Eddington, *The Nature of the Physical World* (Cambridge: Cambridge University Press, 1928), "Introduction."
24. G. E. Moore, "An Autobiography," *The Philosophy of G. E. Moore,* P. A. Schilpp (ed.) (La Salle: Library of Living Philosophers, 1942), p. 14.
25. Cf. Gilbert Ryle, "The World of Science and the Everyday World," *Dilemmas* (Cambridge: Cambridge University Press, 1954), pp. 68–81.
26. Although arrived at independently, this is also the thesis of *Philosophy and the Mirror of Nature,* cited in footnote 9 above. However, from this common thesis we draw opposite conclusions as to the function and future of epistemology, as will be apparent in what follows.

CHAPTER I

THE CARTESIAN LEGACY

THE aim of any theory being to make something more intelligible by explaining it, the purpose of a theory of knowledge is to explicate the knowledge we possess in terms of its essential components and the conditions that make it possible. That we possess various kinds of knowledge, from ordinary facts to the incredible discoveries of the sciences, is hardly to be denied. Moreover, all empirical knowledge, at least, shares certain aspects, such as a derived sensory or empirical content, an imposed conceptual interpretation, and a symbolic formulation. As I proceed, I shall attempt to elucidate the origin and role of each of these aspects of knowledge, deliberately rejecting two interrelated assumptions common to nearly all earlier theories of knowledge: (1) that we must begin from the Cartesian paradigm of an indubitable, mind–dependent content, such as ideas or sense–data; and (2) given the priority of this epistemological position, that any scientific investigation of experience, such as neurophysiology or cognitive psychology, is irrelevant because it is dependent upon a prior epistemological certitude, or more explicitly, that we are precluded from attempting to explain the origin of sensory qualities by neurological processes because they themselves are known only through sensory experi-

1

ences, or even that neurons as perceived, are merely sense data.[1] Given developments in the biological sciences, especially the remarkable progress in neurophysiology made during the past few decades, in contrast to the nearly complete lack of such knowledge prior to the twentieth century, we are no longer compelled to make scientific inquiry subsidiary to purely epistemological considerations.

Accordingly, I shall begin this investigation not with the point of view that man is an immobile, self–existent mind screened from the world by sense impressions, but that he is an active biological creature more highly evolved than, but not evolutionarily discontinuous with, lower biological organisms, and therefore also a product of nature. That he possesses unique capacities, such as the ability to form and to think in abstract concepts, to acquire and creatively use language, to experiment and develop technical instruments to further his inquiries, to invent logical and mathematical calculi to elaborate his observational knowledge, and to institute methods for transmitting this knowledge from one generation to another, does not mean that these various capabilities are independent of his particular physiological and neurological endowments, but just the contrary, as neurophysiological investigations indicate.

Moreover, that our sensory system, like those of other species, has evolved in such a manner as to enable us to adjust to the physical world would seem to indicate that this system conveys some accurate information, however limited and fragmentary, about nature, otherwise it would be difficult to understand our survival. Also, nature is hardly so profligate as to have brought about the evolution of a highly complex central nervous system consisting of billions of nerve cells without its having a reliable function. On the other hand, the invention of instruments like the telescope, microscope, and spectroscope, as well as the experimental discovery of such imperceptible entities as electromagnetic fields, atomic particles, and DNA molecules, indicates that our senses are extremely limited in the range of their response to physical stimuli. And given this stimuli, we are only now beginning to acquire some insight as to how sensory input or information is "encoded" and "processed" by the intricate systems of neurons to produce conscious awareness and appropriate behavioral response.

One should also recognize that, however learned the theoretician, he once was a child and thus all his specialized knowledge had to be acquired from initially rudimentary experiences, which leads to a number of important questions. Is it the case, for example, as Locke (and Aristotle) believed, that the mind of the child is a "blank tablet," an impassive though impressionable material on which the effects of experience leave their impress in the form of simple ideas of sensation, and that all higher or more complex modes of knowledge must be derived from and composed of these sensory elements? That is, while acknowledging certain mental *capacities,* as Locke did, should the *contents* (though ultimately mind–dependent) be attributed to the effects of the outer world on our external senses, which effects, when conveyed to the brain, in some inexplicable manner cause the simple ideas of sensation in our minds, so that these "given" elements are the source of all knowledge? This is the classical empiricist view, which depicts the initial source of knowledge as discrete sensory elements passively derived from experience or nature, which are then combined to form images and complex ideas of external physical objects.

Or is it the case, as Descartes believed, that while certain forms of knowledge comprising self–evident *propositions,* such as logic and mathematics, depend upon some learning experience to be acquired, their origin, as well as their ultimate recognition as self–evident principles, must be attributed to the mind itself? Although the *empirical* contents of knowledge depend upon the senses as their antecedent cause, do other forms of *rational* knowledge originate in the mind itself, so that the mind has not only an innate capacity to respond to acquired contents, as Locke asserted, but is also the *original source* of other forms of knowledge, namely, self–evident rational principles? Such knowledge need not preexist within the mind, but when we begin to think, the mind *derives from itself* these principles, which therefore, cannot be attributed to the senses. This is the classical rationalist view, according to which our most important knowledge is obtained from the nature and function of the mind. This same controversy, that of empiricism versus rationalism, or nature versus nurture (as it is commonly called today), has had its representatives in psychology in the opposition between behaviorists such as James Watson and Clarke Hull, and Gestalt psychologists such as Wolfgang Köhler and Kurt Kaffka, as well as in linguistics between the operant–conditioning program of B. F. Skinner and W. V. Quine, and the innatist–structuralist position of Noam Chomsky.

Neither position is exhaustive, however. Kant, for example, developed a mediating position, acknowledging both a passive aspect of knowledge regarding the initial sensory content, and an active aspect whereby innate forms of perception (space and time) and inherent categories of the understanding (such as substance and causality) are imposed on the "sensory manifold," transforming it into a "synthetic unity of apperception." Thus while the content of experience was thought to be passively acquired, Kant claimed that a form was imposed upon this sensory manifold by the active, synthesizing mind in accordance with its inherent structural categories. Neither perception nor knowledge could be attributed to the sensory contents alone nor to the mind, but was a result of a synthesis of both, with neither component completely independent in function or status.

However, the almost exclusive role attributed to the mind by Kant would be denied today by most scholars. Although Descartes himself presented a detailed account of the function of the central nervous system, his successors progressively accentuated the importance of the mind for experience at the expense of the body. While each acknowledged, at some point, the function of the sense organs, nerves, and brain in conveying "influences" from the outer world to the mind, their main focus of attention was the crucial role of the mind as the locus of experience, since each assumed that the mind was an independently existing entity with its own unique nature and capacities.

Descartes, who performed various optical and physiological experiments prior to writing a book in dioptrics and on human physiology, was concerned with bridging the gap (largely his own formulation) between the material substance composing the bodily machine and the spiritual substance of the soul or mind, especially as he was well aware of their intimate interaction, despite their contradictory attributes. According to his view, the senses originally cause a motion in the "animal spirits" (a subtle fluid) which is conveyed through the nerves to the brain. In the brain, the motions transmitted to the animal spirits surrounding the pineal gland cause the gland to vibrate, which in turn agitates the soul whose "principal seat" is in the gland. The various agitations of the soul, in turn, are the source of such "passions" as feelings, images, perceptions, memories, and thoughts. The following quotation not only exemplified Descartes' model of explanation, it shows his intent to explain such ordinary facts of experience as single vision perception (given that two images are conveyed from the retinae to the brain).

Thus, for example, if we see some animal approach us, the light reflected from its body depicts two images of it, one in each of our eyes, and these two images form two others, by means of the optic nerves, in the interior surface of the brain which faces its cavities; then from there, by means of the animal spirits with which its cavities are filled, these images so radiate towards the little gland which is surrounded by these spirits, that the movement which forms each point of one of the images tends towards the same point of the gland towards which tends the movement which forms the point of the other image, which represents the same part of this animal. By this means the two images which are in the brain form but one upon the gland, which, acting immediately upon the soul, causes it to see the form of this animal.[2]

Similarly, Descartes attempted to explain such experiences as the pain felt in an amputated limb ("the phantom limb"), illusions (caused by the independent movement of the animal spirits in the brain), the correlation of perceptions with external objects, willed notions, and memory. Although Descartes' view is often ridiculed, many of his explanations sound very contemporary, such as his account of memory.

Thus when the soul desires to recollect something, this desire causes the gland, by inclining successively to different sides, to thrust the spirits towards different parts of the brain until they come across that part where the traces left there by the object which we wish to recollect are found. . . . Thus these spirits in coming in contact with these [parts]. . .excite a special movement in the gland which represents the same object to the soul, and causes it to know that it is this which it desired to remember (Article XLII, p. 350; brackets added).

This explanation is not unlike Sir John Eccles' description of the function of the hippocampus in "laying down" memory traces, and of the mind in "recovering" or "reading out" this stored information.[3]

Locke, although himself a trained physician, originally disclaimed any intent to "meddle with the physical consideration of the mind or trouble myself to examine. . .by what motions of our spirits, or alterations of our bodies, we come to have any sensation by our organs,"[4] yet his discussion of the source of our ideas forced him to give some account of their physiological origin. In so doing, he thought of the nerves as "conduits" conveying physical effects to the brain which was depicted as a "presence room" or antechamber of the mind. The following statement is typical of his usual representation of the process.

If, then, external objects be not united to our minds when they produce ideas in it, and yet we perceive these original qualities in such of them as singly fall under our senses, it is evident that some motion must be thence continued by our nerves or animal spirits...to the brains or the seat of sensation, there to produce in our minds the particular ideas we have of them. And since the extension, figure, number, and motion of bodies of an observable bigness, may be perceived at a distance by the sight, it is evident some singly imperceptible bodies must come from them to the eyes, and thereby convey to the brain some motion which produces these ideas which we have of them in us (Bk. II, Ch. VIII, sec. 12).

Although Kant also described "sensibility" as being the "capacity (receptivity) for receiving representations through the mode in which we are affected by objects,"[5] usually he was very careful not to refer to our sense organs as such, since they, like all objects or bodies, were ultimately categorized contents of the mind without independent existence. Even when referring to "outer sense," he did not mean an external sense receptor but the mind's capacity to represent phenomena in an outward, three-dimensional space: "By means of outer sense, a property of our mind, we represent to ourselves objects as outside us. . . ." (B 37). In that respect, Kant was more consistent than the British Empiricists who often began their analysis of perception by describing the role of the sense organs as if they existed independently of the mind when, in fact, their final analysis would have required them to consider the sense organs, as all other objects of immediate perception, as groups of sense qualities, ideas, or impressions. Locke, of course, could think of the sense organs in terms of primary qualities, but he never did so, leaving the reader to make the transition to a different level of analysis. For Berkeley, however, the senses, as all other (supposedly) physical objects, had their being in "being perceived," thus making it difficult to understand how, as ideas, they could have functioned as the mediating cause of sensations. And yet he did presuppose this, as numerous statements indicate: "It is evident to any one who takes a survey of the objects of human knowledge that they are either ideas actually imprinted on the senses, or else such as are perceived by attending to the passions and operations of the mind. . . ."[6] But how could the senses, themselves ideas, receive the imprint of other ideas? A similar inconsistency pervades Hume's analysis of perception.

As indicated, Kant astutely avoided the problem by almost never mentioning the sense organs or the human body, referring to the brain only eleven times, that I have noted, in the entire *Critique of Pure Reason*. But in fairness to the philosophers mentioned, this is a very difficult problem to surmount, a problem analogous to Russell's Paradox of how a class of all entities can avoid including itself as a member of itself. Regarding perception, the paradox pertains to how investigation of the role of the sense organs and the central nervous system in perception can disclose that they are the cause of the sensory qualities which they themselves exhibit within the normal context of perception. More simply, how can the sense organs and the brain be the origin of the sensory qualities (colors and textures) which they manifest to an external observer, such as a neural surgeon, under the usual conditions of perception? Perhaps the enigma can be mitigated, as Russell attempted with his "theory of types," by clearly distinguishing the different contexts of reference or investigation, but it still remains a puzzling phenomena.

Returning to our discussion, the common assumption among classical modern philosophers that all experience and knowledge are mind-dependent would no longer be held by most contemporary philosphers and scientists. In the first place, not having a very explicit conception of what the mind actually is, especially if thought of as a kind of separate spiritual substance within the body, endowing it with complex structures and functions hardly seems to further our understanding of the nature of knowledge—not that knowledge does not presuppose structures and functions, but attributing them to an independently existing mental substance hardly reduces the mystery.

Second, there is considerable empirical evidence to show that the kinds of complex synthesizing activities attributed by Kant to the "transcendental ego" are integrating processes (Sherrington's term) actually performed by the brain. Of the various supporting data, the most impressive pertains to impairment of mental functions caused by brain damage, such as circumscribed or global cerebral lesions, severing of bundles of nerve fibers as in commissural sections, and ablation or surgical removal of portions of the brain. As Howard Gardiner states:

> ... of the varied mirrors which reflect mentation—which illuminate such fundamental capacities as language, perception, memory, artistry—it is brain damage whose study has seemed to me the most rewarding and fascinating.[7]

Norman Geshwind, and more recently, Gardiner (in the work just cited) have described the tragic but intriguing variety of mental impairments attributed to brain lesions caused by strokes. Strokes have resulted in different types of aphasias or speech defects, loss of short–term memory leaving the individual incapable of learning anything new, impairment of motor functions and coordination (such as the ability to write or paint), removal of portions of the visual field, elimination of creative talents, diminished motivation, destruction of one's former sense of a dynamic self–identity, and so forth. Often, a stroke victim will manifest normal behavior in certain respects and yet be completely disabled in others, as the patient who has a normal memory of everything in his past prior to the stroke, but can recall nothing thereafter, or as the composer Maurice Ravel whose sense of pitch, rhythm, and musical taste remained intact after a moderate stroke, but whose abilities to read a score, perform, and compose were seriously affected. There are also fascinating instances of portrait painters whose techniques were markedly changed by a stroke, yet in such a way as to enhance the dramatic quality and aesthetic value of their paintings (pp. 316–334).

The most poignant and revealing account of mental disabilities due to brain damage is that of a Russian soldier, Zasetsky, wounded in the head during the Second World War. The case is documented by the late Russian neuropsychologist A. R. Luria who published the soldier's notebooks which were painstakingly written over a period of twenty-five years.[8] Among the effects caused by irreversible damage to the left parieto–occipital area of Zasetsky's brain are the following: constant buzzing in the head accompanied by frequent aching and dizziness; elimination of the right half of his visual field with the result that he constantly bumped into things and had the horrifying experience of seeing only half his body; the sensation that he had lost parts of his body or that they had become grotesquely distorted in size; disassociation of his stream of consciousness into fragmented and dreamlike contents; inability to distinguish spatial orientations so that he would get lost in the most familiar environment; an initial loss of long–and short–term memory which he partially regained later; an exasperating difficulty in recalling the most familiar words and then in remembering what they meant (when tested, he could not immediately identify his nose and eyes); frustration in understanding the simplest conversation because of difficulty and delay in associating the proper meanings with the words; an incapacity, when reading, to grasp the sense of more than three words at a time; difficulty focusing on an entire word (or even letter) because he could see only the half located in his left visual field; and an inability to comprehend such simple grammatical constructions as "a father's son," even when he understood the meaning of the individual words.

Yet, despite his nightmarish condition, by carefully preselecting words which he jotted down on pieces of paper with infinite patience and perseverance, Zasetsky amazingly compiled a notebook of nearly three thousand pages. For with all his tragic limitations, he had a definite sense of self-awareness, including the dreadful realization of how limited he had become compared with his former capabilities. It was this sense of self that drove him to complete the notebooks.

As it is natural to assume some correlation between such brain lesions and the resultant loss in motor and cognitive abilities, this has provided indispensable evidence for associating mental functions with various areas, structures, or components of the brain. For example, motor aphasia, whereby the individual has lost the ability to speak effectively, although he still can understand spoken language, is centered in the anterior speech center of Broca located in the frontal lobe of the cortex, while the opposite effect, where the individual can speak with normal speed and rhythm but has lost his ability to understand speech, is associated with the posterior speech center of Wernicke located in the temporal lobe. Thus the various areas and parts of the brain, including the four lobes (frontal, parietal, occipital, and temporal) of the two hemispheres of the cerebral cortex, the cerebellum, the hippocampus, the limbic system, and so forth, have been "mapped" in terms of the various conscious functions, both motor and mental, correlated with them.[9]

Wilder Penfield and his associates contributed to this mapping by eliciting responses from fully conscious patients undergoing a cranial operation, while their exposed brains were repeatedly stimulated by electrodes. According to Penfield, these electrode stimulations, though often producing diffuse sensations and "dreamy states," sometimes caused specific and detailed "flashbacks," such as the following:

... a mother told me she was suddenly aware, as my electrode touched the cortex, of being in her kitchen listening to the voice of her little boy who was playing outside in the yard. She was aware of the neighborhood noises, such as passing motor cars, that might mean danger to him.[10]

Although the patients always identified these flashbacks as coming from their own past, they had a passive attitude toward them since they were experienced as being outside their conscious control.

As in the example of the Russian soldier, there is another particularly dramatic instance of this localization, that of commissural section or, as it is more commonly called, the split brain. Some years ago P. W. Sperry and Ronald E. Meyers made the surprising discovery that when the isthmus of nerve tissue called the "corpus callosum" connecting the two hemispheres of the brain is severed, each hemisphere functions independently as a separate brain. The most significant results occurred, however, when not only the corpus callosum but the optic chiasm (the crossover of the optic nerves between the hemispheres) was severed, so that instead of information from both halves of the visual field being transmitted to both hemispheres, as normally happens, information from the right visual field of both eyes was transmitted solely to the left hemisphere, and information from the left visual field was transmitted solely to the right hemisphere.

Although this operation did not alter the patient's personality or general intelligence, it did produce some surprising side effects, indicating that the speech center for most people is located primarily in the dominant *left* hemisphere, while spatial orientation, musical appreciation, and visual pattern recognition are located mainly in the minor *right* hemisphere. For example, when objects were presented to the *left* visual field (and thus information about them transmitted to the nearly non–verbal *right* hemisphere), although the person could recognize the objects (since he followed directions in picking them out), he could not name them. When a word such as 'heart' was projected in the center of his visual field so that the word 'he' fell in his left visual field (and thus was transmitted to the right hemisphere) while 'art' fell in his right visual field (and thus was transmitted to the left hemisphere with the speech center), he answered "art" when asked what word he was seeing.[11] It is apparent from the way the experiment was conducted that the right hemisphere 'recognized' the information transmitted to it, but because the link with the speech center in the left hemisphere had been severed, the person was unable to name the object; when given directions, he could pick out objects in his left visual field but was unable to verbally identify them (indicating, also, that while the right hemisphere cannot produce speech, it can understand it).

The results are particularly interesting because of the unusual glimpse they offer of our dependence on the autonomous functioning of our brains in performing simple mental acts. Asked to *verbally* identify red or green flashes in his left visual field (transmitted to the non–verbal right hemisphere), at first the patient could do so only randomly because the verbal responses given by the left hemisphere were not cued to the visual information presented to the right hemisphere. Soon, however, the erroneous answers could be corrected because the right hemisphere (which received the color information), having heard the wrong answer given by the left hemisphere, signalled the mistake in such a way that the left hemisphere (or person) could correct it. It is significant to note how natural it is to begin talking as if the separate hemispheres were doing the experiencing and the responding for the individual, so that the person seems to be merely a passive medium, as illustrated by the following description of the experimental results:

> What was happening was that the right hemisphere saw the red light and heard the left hemisphere make the guess "green." Knowing that the answer was wrong, the right hemisphere precipitated a frown and a shake of the head, which in turn cued in the left hemisphere to the fact that the answer was wrong and that it had better correct itself![12]

Even more astonishing is the fact that the right hemisphere can recognize objects in the left visual field and direct the left hand to correctly manipulate the objects *without the person being at all conscious* of the activity. If directions for handling objects are presented to the left visual field and thus conveyed only to the right hemisphere, this hemisphere can recognize the objects and direct the left hand (each hemisphere controlling the opposite part of the body) to manipulate the objects according to directions without the person *having any awareness* of what is occurring, unless he turns his head so that the action falls within his right visual field.

These experimental results so impressed Eccles that he concluded that consciousness is associated with only the dominant hemisphere, the minor hemisphere lacking consciousness altogether.[13] However, other investigators such as R. W. Sperry and J. E. Bogen disagree, maintaining that the minor hemisphere is conscious but cannot evince this consciousness because of its inability

to communicate its experiences,[14] raising the intriguing question of
how dependent is conscious awareness on speech. In any case, such
compelling evidence of the lateralization of hemispheric functions
and of the dependence of mental capabilities on various highly
integrated brain structures has convinced many neurophysiologists,
psychologists, and philosophers that instead of thinking of the mind
as existing in some mysterious manner independently of the brain,
we must realize that it is the proper functioning of the brain that
produces normal conscious experiences.

Understanding that higher mental functions are dependent upon
the central nervous system has further important consequences; for
example, the scientists and philosophers of the seventeenth and
eighteenth centuries, in their optical and physiological investigations of visual perception, devised a model of explanation which
contributed significantly to their formulation of the traditional
problem of knowledge—a formulation which can be emended as a
result of recent research in neurophysiology. As we have previously
noted with regard to Descartes and Locke, the underlying paradigm
of the classical modern theories of perception was representational:
the effects conveyed from the senses to the brain producing in the
mind a "representational image" of the external world, at least in
cases of veridical perception. But such an account raises the question of how veridical is such an image. For Descartes, faith in God's
goodness ultimately guaranteed the validity of our perceptions
while Locke maintained (along with the scientists) that we must
assume that certain of our ideas, such as those of primary qualities,
either "truly represent" real physical properties or at least have all
the "agreement needed" for human experience and knowledge,
with Kant denying any agreement between our knowledge of things
as they appear to us and as they are in themselves.

Thus, even at that time there were marked differences of opinion
as to the exactness of the "representational image" produced in the
mind and how it brought about perception of the external world.
Using the camera obscura as a model, Descartes experimented with
eyes removed from the corpses of men and oxen, covering an
excised portion in the backs of the eyes with pieces of paper or egg
shells on which he could see reflected the images of external
objects, to demonstrate how images must be refracted on the retina.
However, he did not rely on this image analogy entirely, asserting
that our external perceptions were not dependent as much on a
"resemblance" between the image and the world, as they were on a

preestablished harmony "ordained by nature" (or God) guaranteeing a certain fidelity between our sensations and their physical causes.

> While this picture [i. e., the retinal image], in thus passing into our head, always retains some degree of resemblance to the objects from which it proceeds, yet we need not think...that it is by means of this resemblance that the picture makes us perceive the objects....Rather we must hold that the movements that go to form the picture, acting immediately on our soul inasmuch as it is united to our body, are so ordained by nature as to give it such sensations.[15] (Brackets added.)

Locke's position was somewhat more complex than Descartes'. Holding a representational view regarding primary qualities, declaring that our "ideas of primary qualities of bodies are resemblances of them," and that such qualities "may be called *real qualities* because they really exist in those bodies," he held a view similar to Descartes' regarding secondary qualities. Like Descartes, he believed that our ideas of secondary qualities were merely "annexed" to the motions producing them, having no resemblance to them at all.

> After the same manner that the ideas of these original [primary] qualities are produced in us, we may conceive that the ideas of secondary qualities are also produced, viz., by the operation of insensible particles on our senses....It being no more impossible to conceive that God should annex such ideas to such motions with which they have no similitude, than that he should annex the idea of pain to the motion of a piece of steel dividing our flesh, with which that idea hath no resemblance (Bk. II, Ch. VIII, sec. 13; brackets added.)

In the final analysis, for Locke as for Descartes, it was "The infinite wise Contriver...[who] hath fitted our senses, faculties, and organs" with just the degree of accuracy to provide for "the conveniences of life...." (Bk. II, Ch. XXIII, sec. 12). Not so easily persuaded were Hume and Kant.

Writing before the era of neurophysiology and electronic computers, these early investigators had no understanding of the present conception of how the central nervous system functions, the main features of which are: (1) the receptor senses (proprioceptors as well as exteroceptors) are transducers which convert the exter-

nal stimulant energy (chemical, electromagnetic, thermal, or mechanical) into the electrical energy of nerve impulses; (2) these
electrical impulses are then converted into electric currents as a
result of the continuous alteration in electrical potential along the
outer membrane of the nerve fibers or axons; (3) these nerve firings
are in turn conducted to multitudes of other nerve cells across
synapses which, by inhibiting or exciting cells with chemical neu-
rotransmitters discharged into the synaptic gap, analyze and modify
the nerve messages; (4) the billions of nerve currents thus produced are conveyed along incredibly complex systems of interconnecting vertical columns of neurons to various interpretive centers
of the brain (such as the visual cortex) where they are decoded and
processed, apparently by specific groups of cells or "cell–assemblies" which respond to very delineated stimuli (such as line orientations); (5) the resultant discharges are integrated by an intricate
cross–over system with data supplied by other components of the
brain, such as the cerebellum and the limbic system (which seems
to be responsible for the emotional tone of our experiences); and
finally, (6) an additional synthesizing process occurs, utilizing the
memory bank presumably stored in the cortex but retrieved by the
hippocampus, before a conscious or unconscious command is signalled to various muscles or glands directing our response to the
external world.[16]

In this astonishingly complex process (especially when considering that it has evolved by chance mutations and selective adaptations) briefly summarized here, there is no question of an image
being conveyed to the brain, as the seventeenth century savants
believed. Not only does the energy level of the nerve impulses not
correspond to the original stimulus energy, but it was discovered by
E. D. Adrian that the nerves are wired as frequency–modulating
systems which respond to the intensity of the stimulus not by
varying the amplitude, but by increasing the frequency of the nerve
impulses. Thus all the diverse information impinging on the receptor cells appears to be transmitted only in two ways, either by
differing frequencies of the nerve firings or by the spatial positions
and interconnections of the neurons. Moreover, while these afferent
nerve fibers convey various data about colors, shapes, odors, and
textures to the brain, no qualitative differences among these fibers
has been discovered, and many cells discharge according to a simple binary on–off system. Such an explanation is hardly consistent
with the notion of a "representational image" transmitted from the

retina to the brain. Rather, from the highly abstract, abbreviated data transmitted to it, the cortex must "reconstitute" the original sensory stimuli.

The above description, which reflects the present state of knowledge and technology as did the seventeenth century conception (and in its own way is probably just as inadequate), depicts the central nervous system as functioning as an elaborate and refined electronic computer. The problem of the representational view disappears since there is no "representational image" to compare to reality. Instead, the question of the adequacy of the system becomes more like that posed by Locke: is the system "fitted. . .to the conveniences of life"? That is, if our central nervous system is not a duplicating machine which should be evaluated in terms of the likeness of its copies, but is a complex data processing instrument which "maps" the external world in order to arouse conscious awareness of possible responses, as well as appropriate unconscious or sympathetic reactions, then the criterion of its effectiveness changes, as does our conception of knowledge. As for knowledge, it becomes much more pragmatic in that the evaluation of the performance of a computer is in terms of the effectiveness of its program in processing data. When one recalls the precision performance of a musician or a fine athlete, and the creative virtuosity of a mathematician, composer, or poet, one cannot but admire nature's extraordinary ingenuity in programming such a marvelously precise, yet creative system. This is especially evident when one considers that computer engineers have been unable to program a computer to simulate riding a bicycle or speak a language, activities easily accomplished by a five-year-old child.

The older paradigm of perception as a copying process is unsupported by recent research in an additional way. As indicated, we do not see a duplication of the retinal image, but *perceive the ordinary world as a result of diverse information conveyed to the brain*. The nature of this stimulation is not that of an image but of linear configurations, surface orientations, angular perspectives, textured densities, and so on. Moreover, rather than the eye functioning as a static camera, constant eye movements or "saccades" are essential for vision because the changing array of retinal patterns supplies necessary cues for interpreting distances, sizes, and orientations. A frog, for instance, will die of starvation if presented only with immobile insects. These constantly changing, abstract, linear schematic data conveyed from the senses are what our brains somehow construct into our ordinary experience of the external world. As graphically depicted by V. B. Mountcastle:

> Each of us believes himself to live directly within the world that surrounds him, to sense its objects and events precisely, to live in real and current time. I assert that these are perceptual illusions, for each of us confronts the world from a brain linked to what is 'out there' by a few million fragile sensory nerve fibres. These are our only informatory channels, our lifelines to reality. These sensory nerve fibres are not high-fidelity recorders, for they accentuate certain stimulus features, neglect others. The central neuron is a story-teller with regard to the afferent nerve fibres; and he is never completely trustworthy, allowing distortions of quality and measure, within a strained but isomorphic spatial relation between "outside" and "inside." Sensation is an abstraction, not a replication, of the real world.[17]

However, having argued that an adequate conception of man requires understanding the indispensable role of the central nervous system in experience, we must recognize how limited this understanding is now, and for the foreseeable future. Such a limitation has been readily admitted by neurophysiologists such as Wilder Penfield:

> Let us consider the brain-mind relationship briefly. . . .It is a boundary which, as some philosophers explain it, does not exist at all. But for the neurophysiologist there is a working boundary that does exist. Physiological methods bring him nearer and nearer to it. But he comes to an impasse, and beyond that impasse no present-day methods can take him.[18]

This "impasse" occurs because we do not have the slightest, remotest idea how the constellation of chemical-electrical nerve impulses eventuate in ordinary conscious experiences! If we start (as any consistent scientific explanation must) *outside* the ordinary context of experience, with various electromagnetic or chemical-physical stimuli affecting the atomic-molecular structure of the different sense organs, which effects are then transmitted as varying frequencies of electrical discharges to the processing centers of the brain, there is no explanation for how the brain constructs the qualitative, three-dimensional, macroscopic world for our conscious awareness. As Locke maintained, "the ideas. . .which we have in our minds can by us be no way deduced from bodily causes. . ." (Bk. IV, Ch. III, sec. 28).

When one listens to a piano recital, for example, whatever sounds, whatever surging or sublime emotions one experiences are due to the effects of the sound waves produced by the impact of the fingers of the pianist on the keyboard conveyed to the ears of the listener, then signalled to the brain where they are integrated and processed to produce musical recognition and sensitivity. Any of a number of brain lesions can destroy forever this marvelous capacity that we all take for granted. Yet we have no more inkling as to how these conscious experiences are produced by cerebral processes than our forebears. Instead of saying that the ideas produced in the mind are "annexed" to the motions in the brain, we are apt to state that our conscious experiences are in some way coterminous with, as well as dependent upon, cerebral processes, but obviously this is no more of an explanation than the former.

To avoid this breach of knowledge, some neurologists and philosophers have adopted the "radical identity theory," claiming that conscious processes are "nothing but" neurological processes.[19] As stated by Steven Rose in his fascinating book, *The Conscious Brain*:

> A large portion of the reductionist/holist dichotomy is based on a semantic confusion as between, say, "mind" and "brain". The interpretation advanced here is a version of what the philosophers describe as the "identity hypothesis"; that "mind" may be defined as the total of brain activity at any given time and "consciousness" as the summation of this total activity. . . ."[20]

If one believes, as I do, that the predominant empirical evidence indicates that mental processes cannot occur independently of their underlying brain structures, then there is a sense in which mind can be identified with the "totality of brain activity" and consciousness equated with the "summation" of this activity. But this hardly constitutes a "definition" or an explanation of the mind or consciousness, being at most a preliminary assumption or working hypothesis, not a solution to the mind–body problem. An electrical engineer could claim that an electric current is the "summation" of the activity of a generator, but he would not then conclude that the electric current is nothing more than the generator; instead, he would go on to explain how a generator produces an electric current and try to describe its properties. (Recently John Eccles has attempted such an explanation regarding the relation of an independent mind to the brain, but I disagree that his interactionist hypothesis is based on "empirical data and is objectively testable."[21] If true, one could conceive of an experimental test of his hypothesis and I do not think this is possible. (Karl Popper, in a personal communication, has agreed with me.)

Consider, for example, the vast difference between the physical properties and theoretical explanation of the neural discharges when a neurophysiologist probes the cortex with an electrode, and the kinds of experiences reported by the patient. They are so completely incommensurate that even if the neurologist could give a description down to the minutest detail of the neural process, this description would not include *anything* pertaining to mental or conscious states, such as pain, thought, feeling, and memory. Any correlation has to be *added externally* on the basis of conscious testimony. There is nothing within the neurological account as such that can explain how or why an individual has the experiences he or she has. Given an established correlation between an experience and the underlying neurological state, we can monitor how changes in the latter affect the former but this always requires going *outside* the physicalistic framework to establish the correlations. Thus it seems to me there is little justification for the prevalent belief among many scientists and philosophers that the scientific framework, as we now envision it, will be sufficient to account for conscious states.[22] However many correlations of conscious experiences with neurological structures can be established, the former's *manifestation as conscious experiences* can hardly be denied, and therefore must be accounted for.

Such gaps or impasses in our knowledge testify to the limitations and inadequacies of our current theoretical explanations, and to the fundamental mysteries that still pervade human existence. Yet Rose, like the positivists and the ordinary language philosophers referred to in the *Introduction,* professes to find no intractable mysteries in life, only particular puzzles that will be cleared up by further investigation.

> Essentially, what I am saying is that there are no properties of the brain which cannot be analyzed, defined, explained and interpreted in terms of the biological mechanisms which are known to operate in other systems. There are, to put it bluntly, *no mysteries.* There are puzzles and problems. There will surely be major surprises to come. There are vast murky areas where knowledge is at best sketchy. But it does not seem to me that there are insoluble paradoxes or major explanatory principles lacking in order to achieve the interpretation

of brain and behavior, where we cannot now begin to provide the outline of a plausible mechanism (p. 275).

This optimism pertains to expected progress in neurobiological explanations, ignoring the mind–brain impasse, since Rose believes, in contrast to Penfield, that the impasse has been surmounted by the 'hypothesis' (which is really only a declaration of faith) that consciousness is "identical with" brain processes. In fairness to Rose, he does admit to a certain naivety in his position: "It may be that this is a statement full of naive hubris, but such has been the development of biology in the last decades that I make it with no apology, and only faint doubts that I may be proved wrong" (p. 275).

From my experience, having seen so many recent dogmas fall by the wayside, such as the principles of verification and falsification, the indubitability of protocol statements, the sense–data theory, the picture theory of language, and the absolute distinction between analytic and synthetic statements, and recalling how physicists in the latter part of the nineteenth century were confident they were close to solving all physical problems, except for "two small clouds on the horizon" (the negative results of the Michelson–Morley experiments and the failure of the Rayleigh–Jeans law to predict the distribution of blackbody radiation), I am insulated against such electrifying claims. Moreover since any complete neurophysiological explanation would depend to some extent upon the theory of quantum mechanics, which is fraught with momentous "paradoxes," as atomic physicists admit, such hubris does seem to be naive, even from the scientific standpoint.

Nor, in my opinion, can the mind–body problem merely be brushed aside by attributing it, as Rose does, to a "semantic confusion between. . .'mind' and 'brain'," or as Wittgenstein did—setting the precedent—to a "logical sleight-of-hand": "The feeling of an unbridgeable gulf between consciousness and brain-process. . .is accompanied by slight giddiness,—which occurs when we are performing a piece of logical sleight-of-hand. . . ."[23]

But is the extraordinary contrast between our conscious experience and the neurological description of the bases of this experience merely to be dismissed as a "slight giddiness" produced, as it were, by a "logical sleight-of-hand"? Compare Wittgenstein's philosophical assessment of the problem with that of Eccles who has devoted his entire professional life to unraveling the complexities of the vast network of neurologial processes underlying our conscious experience.

> Is it not true that the most common of our experiences are accepted without any appreciation of their tremendous mystery? Are we not still like children in our outlook on our experiences of conscious life, accepting them and only rarely pausing to contemplate and appreciate the wonder of conscious experiences? For example, vision gives us from instant to instant a three-dimensional picture of an external world and builds into that picture such qualities as brightness and color, which exist only in perceptions developed as a consequence of brain action. Of course we now recognize physical counterparts of these perceptual experiences, such as the intensity of the radiating source and the wave lengths of its emitted radiation; nevertheless, the perceptions themselves arise in some quite unknown manner out of the coded information conveyed from the retina to the brain. Perhaps it is easier still to appreciate the miraculous transformation that occurs in hearing—from mere congeries of pressure waves in the atmosphere to sound with tone and harmony and melody.[24]

According to the accepted scientific explanation, the interaction of the human organism (itself a biomolecular structure) with the microstructure of the physical world produces a conscious awareness of a macroscopic world qualitatively different from either the substructure of nature or the patterns of electrical discharges in the brain. But, despite the qualitative difference between the macroscopic and the microscopic domains, the former provides the evidence for the latter, while our knowledge of the latter explains many of the properties and occurrences, otherwise inexplicable, of the former.

This paradoxical situation is not produced either by a "semantic confusion" or by a "logical sleight-of-hand"; it is due, rather, as almost all past philosophers and scientists realized, to the incredible difference between the scientific *conception* and the ordinary *experience* of the world. As an expedient, one can avoid the paradox by remaining within either context, that of everyday experience and ordinary language, as Moore, Ryle, and other ordinary language philosophers did, or that of the scientific framework, as Rose and philosophical reductionists attempt to do. But as soon as we try to give a complete, coherent account of experience, the hiatus or impasse in our explanations becomes apparent. Wittgenstein's analysis is typical of an analytic philosopher's attempt to explain away our ignorance of some phenomenon by attributing the perplexity to a logical, semantic, or conceptual confusion.

Due to the pervasive effect of Cartesian dualism, Western man since the seventeenth century has tended to think of himself as a kind of independent spiritual substance or soul, as if the body were an incidental appendage to the mind. Now, however, as a result of the influence of the physical and neurological sciences, there is a tendency to regard man as no more than overt behavioral responses directed by a "central state" computer or brain. But just as it was a conceptual distortion to conceive of human beings as "minds annexed to bodies" or as "ghosts ensconced in a machine" (Ryle's graphic terminology), it is a reverse conceptual contortion to think of human beings as mindless computers or biological robots.[25] The fact that our conscious abilities depend upon a genetically programmed support system does not warrant denying the efficacy of these abilities. Just as biologists recognize that man's varied physical adaptations to the world have depended upon his uniquely evolved brain, so neurophysiologists should recognize that man's considerable cultural creations have depended upon his emergent consciousness: that consciousness, rather than proven biologically superfluous by neurological research, is a necessary condition of man's unique acquisition and use of language, and therefore of his cultural evolution. As Popper has aptly stated: "The emergence of full consciousness, capable of self–reflection, which seems to be linked to the human brain and to the descriptive function of language, is indeed one of the greatest of miracles."[26]

Nevertheless, in an understandable effort to eliminate the discontinuities inherent in dualism, various attempts have been made to justify a purely physicalistic framework. In addition to the reductionist position discussed above, psychological and philosophical behaviorists have tried to reduce or assimilate inner conscious states to external behavioral manifestations. Following the suggestions of Wittgenstein in the *Philosophical Investigations,* Ryle, for example, formulated a dispositional concept of the mind according to which subjective experiences were reinterpreted as dispositions to overt behavior. To accomplish this, he had to minimize the seemingly "privileged access" one has to such private experiences as dreams, hallucinations, and images because, as usually interpreted, they attest to a domain of inner conscious states inaccessible to public scrutiny. As an example of his approach, Ryle suggested that rather than talking of "seeing images" one should say a person is having an experience similar to what he would have if he were seeing an actual external object.[27]

Such attempts, however, have always seemed to me to be contradicted by readily acknowledged experiences and undeniable empirical data. As for images, there is the evidence of eidectic imagery and nearly total recall, as well as vivid dreams and hallucinations. Eidetic images can be so predominant that they resemble a photographic replica of a particular scene, having an authenticity and independent spatial location similar to normal perceptions, along with the capacity of blotting out an area of the visual field. In addition, there is the extraordinary example of the mnemonist described by A. R. Luria who's recall was so nearly total that after a period of thirty years he could still recite, as if he were reading it, a long list of unrelated words he had once seen in a language completely foreign to him.[28] What is striking about these experiences is that the images, although subjective, appear to the individual as objective data, very much like normal perceptions.

As for dreams, though it is not yet certain what purpose they serve, they seem to be necessary for our mental stability. If subjects are awakened from deep (paradoxical) sleep, so as to interrupt the dream cycle, they may become psychotic in a period of two or three days. Then there are the examples of drug–induced hallucinations, some of which are so vivid, realistic, and terrifying that they supercede the ordinary world of perception, inducing a few individuals to jump out of windows or to run in front of automobiles, with fatal consequences. As Ulric Neisser states, those thinkers (such as Ryle) who oppose images assume "that only poetic fantasy allowed one to speak of 'seeing' in connection with. . .mental imagery;" empirically, however, this is not the case. Some people

> were quite ready to describe their mental imagery in terms normally applied to perception. . .some psychopathological states can endow images with such a compelling quality that they dominate the patient's experience. Students of perception have often disregarded dreams and phantasms, considering them "hallucinatory" and thus irrelevant to normal seeing. However, this is a difficult position to defend either logically or empirically.[29]

While Ryle is perhaps a special case (although writing a book on the "concept of mind" he boasted of never having studied psychology[30]), nevertheless, this is an excellent example of firm empirical evidence contradicting philosophical arguments advanced in an *a priori* manner to support a particular methodological or ontological position. Even if, following Ryle's suggestion, we did describe the experience of seeing subjective images as similar to what we would experience if we were perceiving an actual object, I fail to see how this eliminates the problem of the nature and status of the *content* of what one is experiencing. Gordon Taylor makes the interesting observation that thinkers fit into two categories, visualizers or verbalizers, with analytic philosophers, especially, falling into the second category, which he thinks explains the apparent vacuity of much of analytic philosophy (cf. pp. 214-215, 303).

A position similar in intent to reductive materialism and behaviorism, in that it too would recognize only the theoretical framework of science as having any validity, is "eliminative physicalism." However, to avoid the difficult problem of showing how the referents of neurological terms and those referring to mental states and sensory qualities can be identical, with the latter reduced to the former, eliminative physicalists maintain that as the scientific picture of the world becomes more complete, the references to those aspects of experience that seem impervious to scientific interpretation or assimilation will be discarded, as our previous beliefs in spirits, daemons, and souls. While this position has also been defended by Quine and Feyerabend, among others,[31] the most imaginative and persuasive presentation is by Paul Churchland.

Accepting the revolutionary thesis that has been widely adopted recently—owing to the cumulative arguments of Wittgenstein, Hanson, Popper, Sellars, Quine, and Feyerabend—that there is no indubitable given content of experience or incorrigible foundation of knowledge, Churchland argues that even a viewpoint as seemingly self-evident and irrefutable as our immediate awareness of our own conscious states still employs a theoretical interpretation (which must be partially true, since the ancient Greeks, especially Aristotle, did not have a conception of the mind analogous to Cartesian dualism), and therefore can be altered or discarded. There are at least three aspects to this argument: (1) all perceptions, introspections, and experiences, however apparently immediate and indubitable, are "theory-laden" or "framework dependent," and therefore corrigible; (2) not just our theories but also our ordinary or immediate perceptions have at times proven to be systematically misleading, as was true of pre-Copernican and pre-Einsteinian observations; hence, (3) as science progresses, no beliefs based on perceptions, observations, or introspections should be considered

immune from possible revision or rejection. As a result, powerful
(scientific) theories can be expected to override our immediate
perceptions, common sense beliefs, and entrenched linguistic conventions—contrary to the view of ordinary language philosophers
such as G. E. Moore and P. F. Strawson. As Churchland states:

> Our perceptual judgments can no longer be assigned any privileged
> status as independent and theory-neutral arbiters of what there is in
> the world. Excellence of theory emerges as the fundamental measure
> of all ontology. The function of science, therefore, is to provide us
> with a superior and (in the long run) perhaps profoundly different
> conception of the world, *even at the perceptual level.*[32]

Churchland goes on to provide ingenious examples of people
who possess different sense modalities (creatures like ourselves in
all respects except their eyes respond to infrared rays in such a way
that they can *visually* perceive the temperatures of objects, rather
than their colors, just as we *tactually* feel heat) or who hold
different theoretical frameworks from ours (people whose experiences of temperature are explained in terms of a *direct perception*
of a compressible fluid called caloric) to show that: (1) the "meaning" of a term. . .is not determined by the intrinsic quality of whatever sensation happens to prompt its observational use, but by the
network of assumptions/beliefs/principals in which it figures. . . ."
(p. 15), and therefore, (2) that a conceptual framework can be so
embedded in or intrinsic to one's perceptions and experiences that
it determines what we perceive, without our realizing that it is a
theoretical framework at all!

The conclusion derived from this, *mutatis mutandis,* is that the
belief that we have an incorrigible private access to inner conscious
states—defended by philosophers such as Kripke[33]—with indubitable intrinsic properties incommensurate with the intersubjective,
physicalistic framework of science may be mistaken, so that in the
future we might abandon this common sense conception of
ourselves.

> If our conceptual framework for P-states ["the Person-theory of hu-
> mans"] is an empirical theory, then it is possible, at the limit, that said
> theory be wholly false, that there are no such things as P-states, that
> *all* of our introspective judgements have been systematically false by
> reason of presupposing a false background theory. This is just con-
> ceivable, on our view, but inconceivable if introspection provides an
> infallible view into our own natures (p. 96; brackets added.)

While this conclusion is stated somewhat hypothetically, Churchland later adds that "the P-theory apears as a dead-end approach to the problem of human nature," that it "will prove too confused and superficial to find any natural reconception or refor-mulation within the framework of a truly adequate theory of our internal activity," and consequently, confesses to "a strong inclina-tion towards eliminative materialism. . . ." (pp. 115-116).

Like the reductive materialists, Churchland envisions a futuristic society that has been so acculturated by developments in science that its citizens have replaced all references to conscious, sensory experiences with their correlated physical and neurological causes, offering an arresting description of what such people would experi-ence at the seashore.

> These people do not sit on the beach and listen to the steady roar of the pounding surf. They sit on the beach and listen to the aperiodic atmospheric compression waves produced as the coherent energy of the ocean waves is audibly redistributed in the chaotic turbulence of the shallows. . . .They do not observe the western sky redden as the Sun sets. They observe the wavelength distribution of incoming solar radiation shift towards the longer wavelengths. . . .They do not feel common objects grow cooler with the onset of darkness, nor observe the dew forming on every surface. They feel the molecular KE of common aggregates dwindle with the new uncompensated radiation of their energy starwards, and they observe the accretion of reassoci-ated atmospheric H_2O molecules as their KE is lost to the now more quiescent aggregates with which they collide. . . .They do not warm themselves next to the fire and gaze at the flickering flames. They absorb some EM energy in the 10^{-5}/m range emitted by the highly exothermic oxidation reaction, and observe the turbulences in the thermally incandescent river of molecules forced upwards by the denser atmosphere surrounding (pp. 29-30).

Although one might think that this description inadvertently confounds "listens to," "observes," "feels," and "warms" with the causal explanations of these experiences, Churchland certainly means to be taken literally. I wonder, however, if advocates of this position are fully aware of its implications; for not only would it eradicate psycho-physical dualism, it would also eliminate all of those glorious manifestations of nature that we so enjoy—all sen-sory qualities such as sounds, colors, fragrances, as well as sensa-tions or experiences like delight, warmth, and contentment. And, having denuded nature of its rich outer garb, what would the effect be on aesthetic creativity, such as poetry (Walt Whitman), painting

(Winslow Homer), and music (Claude Debussy)? Would artists be inspired to represent neuronal discharges?

While appealing as regards its scientific unity and simplicity, eliminative physicalism (and reductive materialism and behaviorism) raises serious problems. If, as argued, there are no intrinsic qualities of perception and introspection that can serve as unassailable indications of real features of the world (as may well be the case), and all sensory observations are theory–laden, then *any* interpretation of experience *is* subject to change by more "powerful theories." But this raises the question, what constitutes a more powerful theory? It is now evident that the heliocentric is a more powerful theory than the geocentric and that relativity 'theory' is superior to Newtonian mechanics, in that the former theories were found to provide better predictions and explanations of empirical phenomena, along with showing how the superceded theories could be derived from them within limiting conditions. It is not evident, however, in what way eliminative or reductive materialism is a more powerful theory of ourselves as total human beings—in contrast to a one–dimensional conception of man as scientist.

In what way do physicalistic theories, which deny the reality of sentient experiences, lead to a fuller or more adequate conception of mankind? If adopted, what would the effects be on our understanding of ourselves as sensitive, aesthetic, morally responsible creatures (as well as scientific researchers)? Would moral responsibility be consistent with the reduction of feelings of obligation or guilt, as well as intentional motives, to certain brain functions? Could one hold brain states responsible for one's actions or accomplishments? Also, while physicalism is more consistent with neurophysiology and with a form of tough–minded scientism characteristic of some Western thought, dualism would seem to be more compatible with Western religion and Eastern mysticism, with Freudian psychoanalysis and cognitive psychology, as well as with the claims of parapsychology. Although we do *not* know to what extent these latter experiences and their interpretations are veridical, their *possible* significance underscores that what constitutes a more powerful theory is not an easy question to answer. It could be argued that adopting a physicalistic view of man (as do most neurophysiologists), on the grounds that at present it is more consistent with the physical sciences, impedes an appreciation and investigation of these more arcane forms of experience which, *if* they should have some validity, would bring about a fundamental revolution in our thinking about ourselves and the physical world.[34] None of us, of course, knows what the final outcome will be; we are merely conjecturing as to the possible eventualities of future developments. While modern science, as Churchland has persuasively argued, can exploit the information inherent in our ordinary sensations or perceptions in radically novel ways to bring about both "an expansion of perceptual consciousness" as well as "profoundly different conceptions of the world," we should bear in mind that there could be features of experience or of nature which cannot be accounted for (at least at a particular stage in the development of science), but for that reason should not be "eliminated." As imperfect as the present dualistic, inner–outer picture of ourselves is, the prospect of a fuller explication of this view in terms of unforeseeable empirical discoveries, scientific advances, and conceptual innovations seems to me more justified than just denying consciousness altogether because it cannot be assimilated into the current scientific framework.

A more promising position has been proposed recently by John Searles. I find it more promising because it accepts the irreducibility of conscious states, attempting to show how they might be fitted into the physicalistic scientific framework, and because it presupposes a conception of autonomous levels similar to the position of "contextual realism" that will be described in the final chapter of this book. Searles defends a staunchly realistic interpretation of mental states, claiming that "there really are such things as intrinsic mental phenomena which cannot be reduced to something else or eliminated by some kind of re–definition,"[35] while also attempting to show how these "mental states are both *caused by* the operations of the brain and *realized in* the structure of the brain. . . ." (p. 265).

His interpretation depends upon recognizing (as I advocate in the final chapter) that while experienced phenomena such as water or tables consist of basic internal elements with their inherent structures, and exist relative to more pervasive background conditions, as long as these inner components and external conditions persist, *the experienced phenomena really do exist with their manifest qualities and properties.* Using water as an example, Searles maintains that it is the combined motions of the H_2O molecules at the microlevel that account for its surface qualities and macroscopic properties, such as fluidity, solvency, transparency, and color. Yet while these surface features of the water are "caused by" the motions of the H_2O molecules and "realized in" their atomic-molecular structures, nonetheless, they exist on their own level with their attendant causal effects, such as flowing, wetting, dissolving, and quenching. Analogously, according to his position of "biological naturalism,"

> mental phenomena are biologically based: they are both caused by the operations of the brain and realized in the structure of the brain. On this view, consciousness and Intentionality are as much a part of human biology as digestion or the circulation of the blood. It is an *objective* fact about the world that it contains certain systems, viz., brains with *subjective* mental states, and it is a *physical* fact about such systems that they have *mental* features. The correct solution to the "mind-body problem" lies not in denying the reality of mental phenomena, but in properly appreciating their biological nature (p. ix).

Initially, Searles' interpretation is what most of us tend to believe if we are scientific realists and accept the reality of such microparticles as atoms and molecules, but nonetheless are unwilling to renounce macroscopic states and phenomenal qualities. Yet I think Searles is somewhat sanguine in thinking this removes the mystery: "the logical nature of the *kinds* of relations between the mind and the brain does not seem to me in that way at all mysterious or incomprehensible" (p. 267). He says, it "remains an open empirical question how these higher-level states are realized in and caused by the operations of the brain...." (p. 271). But as I have argued elsewhere, [36] it is precisely at this point that explanations in science seem to break down. Although we can analyze such compounds as water, salt, and hydrochloric acid into their atomic elements, and can explain their molecular composition in terms of atomic structures and chemical bonds, scientists cannot explain how or why these structures produce the different macroscopic properties they do; why combinations of NaCl have the sensory qualities of salt and HCl those of an acid. Insofar as we are able to explain the physical state of a compound (its being a solid, liquid, or gas) by means of its molecular motions, and account for its chemical makeup (such as being an acid) in terms of such atomic properties as possessing free hydrogen ions, one could reply that it *is* possible *even now* to explain macroscopic features in terms of microstates and processes. But how can we account for the colors, tastes, and odors of the compounds, and for the forms of their *perceptual* states? These qualities seem to arise as a result of the effect of the compounds on some organism's sensory receptor, in which case we are again faced with the question of how these occur or emerge in the brain.

Searles is not unaware of this problem, but I think his answers avoid direct confrontation with it. Regarding the example of water, he asks the crucial question:

> So if one asked, "How can there be a causal relation between the molecular behavior and the liquidity if the same stuff is both liquid and a collection of molecules?", the answer is that there can be causal relations between phenomena at different levels in the very same underlying stuff. . . .To generalize. . .we might say that two phenomena can be related by both causation and realization provided that they are so at different levels of description (p. 266).

The difficulty is that Searles states as a fact what we would like to have clarified or explained!

This is true also as to how mental states, which he claims "are both *caused by* the operations of the brain and *realized in* the structure of the brain," are actually *produced*. We can agree that the "liquidity of water is not to be found at the level of the individual molecule," and that "visual perception. . .[is not] to be found at the level of the individual neuron or synapse" (p. 268), yet one wonders how and where they do originate. The assertion that they are caused by and realized in the microstructures does not explain how this is possible. One of his statements particularly points up this problem.

> Suppose we had a perfect science of the brain, so that we knew in detail how brain functions *produced* mental states and events. If we had a perfect knowledge of how the brain *produced,* for example, thirst or visual experiences, we would have no hesitation in assigning these experiences locations in the brain, if the evidence warranted such assignments (pp. 270–271; italics added.)

But this assumes the very question at hand: in what sense "a perfect science of the brain" could explain how brain activity "produces" conscious states. If my own evaluation is correct, there is absolutely nothing in our present knowledge of neurophysiology to indicate what such an explanation would be. Searles himself is aware of this, insofar as he later acknowledges that entirely new scientific principles may be required for such an explanation.

For all we know the type of realizations that. . .[mental] states have in
the brain may be describable at a much higher functional level than
that of the specific biochemistry of the neurons involved. My own
speculation, and at the present state of our knowledge of neurophysiology it can only be a speculation, is that if we come to
understand the operation of the brain in producing. . .[mental
states], it is likely to be on principles that are quite different from
those we now employ, as different as the principles of quantum
mechanics are from the principles of Newtonian mechanics; but any
principles, to give us an adequate account of the brain, will have to
recognize the reality of, and explain the causal capacities of,
the. . .[mental states] of the brain (p. 272). [Within the brackets
"mental states" have been substituted for the deleted words, "Intentional" or "Intentionality".]

But the more intriguing question is not whether radically new
principles would be required to provide a scientific explanation of
how mental states and phenomenal qualities are produced by physiological or physical processes (which I think is obvious), but
whether science as we know it is capable of solving these kinds of
problems![37] At this point we just do not know.

Thus I suggest we follow the lead of quantum mechanics, where
physicists use complementary concepts such as particle and wave,
energy and time, and velocity and momentum, acknowledging that
the experimental situation determines which of these mutually
exclusive but complementary forms will be manifested, and employ
the concept of "cerebral–conscious states" in recognition of these
indispensable, mutually exclusive, but complementary realities. As
in quantum mechanics, whether we encounter mental or neural
processes seems to depend upon the conditions of experience, the
neurophysiologist having access only to neural discharges (as Penfield acknowledged) while the patient normallly undergoes conscious experiences (as opposed to their underlying cause).
Furthermore, if we accept the causal interdependence but autonomous status and causal efficacy of each of these levels of activity, as
Searles also proposes, we can utilize the powerful tools of description and explanation of both the brain sciences and cognitive psychology or phenomenology.

Before concluding this chapter, I want to mention a different
approach to the legacy of Cartesian dualism, that of Richard Rorty in
Philosophy and the Mirror of Nature. Recommending that philosophers "abandon" the identity–theory of either reductive materialism
or eliminative physicalism (which he previously had defended),

Rorty presents an historical exegesis of Cartesianism and its effects on European philosophy to show that although the tradition has been the source of the major developments and schools of modern philosophy, it now has "lost its vitality" and is "outworn."[38] While it is impossible to do justice to Rorty in a brief space, I nevertheless would like to mention a few points on which I agree and disagree, especially as his book has been widely discussed. In my opinion, Rorty is correct in attributing the Cartesian paradigm of an inner private consciousness to Decartes' conviction that knowledge of the contents of this inner consciousness, in contrast to knowledge of the external world, is "indubitable." As I pointed out in the Introduction, a crucial component of the controversy attending Galileo's defense of the Copernican view was whether we can trust our senses or whether they deceive us. For if the Copernican hypothesis were true, then our ordinary sensory perceptions (as opposed to Galileo's astronomical observations) would be systematically misleading, raising the question whether there is a more secure foundation of knowledge. Contrary to Rorty's assertion (cf. p. 55), Descartes does answer the question "explicitly," maintaining that while we can doubt the validity of any perceptual judgment, we cannot doubt that we are having the experience on which the judgment is based: that in so far as we undergo the experience without claiming that it is representative of anything, it cannot be false but must be indubitable, from which he inferred that such experiences are epistemically and ontologically independent of physical reality.[39]

From this conviction, coupled with the prevalent belief that what we directly sense or perceive is not the external world but the effects of this world on our sense organs as conveyed to the brain, which then in some inexplicable way results in representative sensations or ideas in the soul or mind, Descartes fathered the notion of the mind as an inner "glassy essence" or mental "mirror" in which the external world is somehow reflected. Accepting Descartes' paradigm of the epistemic and ontological priority of conscious experience, Locke in turn defined an idea as "Whatsoever the mind perceives in itself, or is the immediate object of perception, thought, or understanding" (Bk. II, Ch. VIII, sec. 8), thus imposing the "veil of ideas" between ourselves and the external world. This resulted in "the epistemic turn in philosophy," with the problem of knowledge and the philosophy of mind becoming major preoccupations of modern philosophers. This analysis by Rorty is, I believe, essentially correct.

What I disagree with is his conviction that because we can trace dualism to the seventeenth century revolution in philosophy initiated by Descartes it is merely "optional;" and that since on Rorty's Wittgensteinian view "an intuition is never anything more or less than familiarity with a language–game" (pp. 34, 46, 136), appeals to intuitive or experiential evidence in support of dualism are nothing more than preferences for a particular form of language. Although I pointed out earlier in this chapter that the representational theory of knowledge was influenced by certain empirical considerations (such as the camera obscura model of perception and the notion of the nervous system functioning as a hydraulically operated duplicating machine), which seems outmoded in terms of what we know today about the neurological functioning of the brain, I do not believe that the cluster of epistemological problems generated by the modern revolution in science is thereby outmoded or "outworn".

In my judgment, the "modern epistemological turn" did not have its *primary* origin in Descartes' view of an inner glassy essence, but that *both* originated in the revolutionary scientific developments of the seventeenth and eighteenth centuries: in the replacement of the older finite, geocentric conception of the universe with the new heliocentric, and subsequently infinite, theory of the universe; in the shift from the ancient Aristotelian organismic cosmology to the modern mechanistic model conceived by Descartes but fundamentally constructed by Newton; and in the rejection of Aristotle's direct realistic theory of perception for a representational view owing to the acceptance of the atomic or corpuscular theory of matter (with the Galileo–Boyle–Locke distinction between primary and secondary qualities), the renewed empirical investigations of the physiology of perception by Felix Plater, Joseph Duverney, Marin Mersenne, and Descartes himself, the optical investigations of Kepler and Newton, and Roemer's proof of the finite velocity of light in 1675. These developments undermined fundamental presuppositions, challenging ordinary convictions, entrenched concepts, and established linguistic conventions. It was to these cosmic changes that the savants of seventeenth and eighteenth century philosophy were responding.

Since we live in a century marked by even greater scientific achievements, the epistemological challenge to comprehend, interpret, and assimilate these develoments by revising or reconstructing our conceptual–linguistic frameworks is equally great. As Rorty acknowledges, his book, "like the writings of the philosophers I most admire, is therapeutic rather than constructive" (p. 7). In opposition to the founders of modern philosophy who were attempting "constructive" solutions to the profound empirical and theoretical problems they confronted, and in contrast to Dewey (one of the three twentieth century philosophers Rorty particularly admires) who considered the role of philosophy that of "reconstructing" traditional beliefs to bring them in harmony with ongoing developments in science, Rorty has been so influenced by the analytic tradition that he still views philosophy as essentially therapeutic—as not attempting to *resolve*, but as claiming to *dissolve* traditional philosophical problems. Ironically, however, he applies this method to the techniques and results of analytic philosophy itself, with the not surprising result that in the end there is not much left to philosophy or philosophizing. Thus the self-destructive tendency inherent in the much acclaimed "linguistic turn" of analytic philosophy (particularly evident in positivism and ordinary language philosophy) has come full circle!

Notes to Chapter I

1. Cf. C. I. Lewis, *Mind And The World Order* (New York: Dover Publications, 1929), p. 56; H. H. Price, *Perception* (London: Methuen, 1983), p. 2; and A. J. Ayer, *Foundations of Empirical Knowledge* (London: Macmillan & Co., 1951), p. 2. W. V. Quine makes the same point in *Ontological Relativity and Other Essays* (New York: Columbia University Press, 1969), pp. 84–85.

2. René Descartes, "The Passions of the Soul," Article XXXV. E. S. Haldane and G. R. T. Ross, *The Philosophical Works of Descartes,* Vol. I (Cambridge: Cambridge University Press, 1967), pp. 347–348.

3. Cf. Karl R. Popper and John C. Eccles, *The Self and Its Brain* (New York: Springer International, 1977), ch. E8.

4. John Locke, *An Essay Concerning Human Understanding,* Bk. I, ch. I, sec. 2. The following references to Locke are to the *Essay.*

5. Immanuel Kant, *Critique of Pure Reason,* trans. by N. K. Smith (London: Macmillan & Co., 1929), "Transcendental Aesthetic," sec. I, par. 2, B 34.

6. George Berkeley, *A Treatise Concerning The Principles Of Human Knowledge,* Part I, sec. 1.

7. Howard Gardner, *The Shattered Mind* (New York: Vintage Books, 1974), p. xi.

8. Cf. A. R. Luria, *The Man with a Shattered World,* trans. by Lynn Solotaroff (New York: Basic Books, 1972).

9. Cf. Karl R. Popper and John C. Eccles, *op. cit.,* pp. 227–406.

10. Wilder Penfield, *The Mystery of the Mind* (Princeton: Princeton University Press, 1975), pp. 21–22.

11. Following precedent, I shall use double quotation marks around written and verbal quotations and single quotation marks to indicate that I am referring to a word or phrase in contrast to its referent (the word 'tree' versus its referent tree), and occasionally to indicate that the word or phrase has an unusual meanings or is being used in a peculiar way.

12. Michael S. Gazzaniga, "The Split Brain in Man," in R. Held and W. Richards (eds.), *Perception: Mechanisms and Models* (San Francisco: W. H. Freeman & Co., 1950), p. 32. Also see Charles E. Marks, *Commissurotomy Consciousness and Unity of Mind* (Cambridge: MIT Press, 1981).

13. Cf. John C. Eccles, *Facing Reality* (New York: Springer–Verlag, 1970), pp. 78–79.

14. Cf. Karl R. Popper and John C. Eccles, *op. cit.,* ch. E5.

15. A. C. Crombie, "Early Concepts Of The Senses And The Mind," in R. Held and W. Richards (eds.), *op. cit.,* pp. 15–16.
16. Cf. Karl R. Popper and John C. Eccles,, *op. cit.,* ch. E8. Also see A. G. Karczmar and J. C. Eccles (eds.), *Brain and Human Behavior* (New York: Springer–Verlag, 1968).
17. V. B. Mountcastle, "The view from within: Pathways to the study of perception," *Johns Hopkins Medical Journal,* 136, pp. 109-131. Quoted from Karl R. Popper and John C. Eccles, *op. cit.,* p. 253.
18. Wilder Penfield and Lamar Roberts, *Speech and Brain Mechanisms* (Princeton: Princeton University Press, 1959), p. 8.
19. For the views of such radical reductionists as J. J. C. Smart, U. T. Place, and D. M. Armstrong, see their articles in C. V. Borst (ed.), *The Mind/Brain Identity Theory* (London: Macmillan & Co., 1970).
20. Steven Rose, *The Conscious Brain* (New York: Alfred A. Knopf, 1975), p. 273.
21. Karl R. Popper and John C. Eccles, *op. cit.,* p. 375.
22. Cf. Gordon R. Taylor, *The Natural History of the Mind* (New York: Penguin Books), pp. 296, 305, 315, 319. I regret not having read Gordon Taylor's excellent book before I had completed my manuscript, preventing me from making references to it except in the editor's copy. However, I was delighted to find such an independent convergence of views, as regards both "the emptiness of purely verbal analysis (p. 303)" and his aproach to and final assessment of the mind-body problem, as in need of "a major new conceptual system" (p. 305).
23. Ludwig Wittgenstein, *Philosophical Investigations,* trans. by G. E. M. Anscombe (Oxford: Basil Blackwell, 1952), Part I, sec. 412.
24. John C. Eccles, *op. cit.,* pp. 1–2.
25. Cf. A. R. Anderson (ed.), *Minds and Machines* (Englewood Cliffs N.J.: Prentice-Hall, 1964); also John Haugeland, *Artificial Intelligence* (Cambridge: MIT Press, 1985).
26. Carl R. Popper and John C. Eccles, *op. cit.,* p. 129.
27. Cf. Gilbert Ryle, *The Concept of Mind* (London: Hutchinson's University Library, 1949), p. 220.
28. Cf. A. R. Luria, *The Mind Of A Mnemonist,* trans. by L. Solotaroff (New York: Basic Books, 1968).
29. Ulric Neisser, "The Processes of Vision," in R. Held and W. Richards (eds.), *op. cit.,* pp. 256–257.
30. Cf. Bryon Magee, *Modern British Philosophy* (New York: St. Martin's Press, 1971), pp. 102–103.
31. Cf. W. V. Quine, *Ontological Relativity and Other Essays, op. cit.,* pp. 82–83; also Paul Feyerabend, "Materialism and the Mind–Body Problem," in C. V. Borst (ed.), *op. cit.,* pp. 142–156.
32. Paul M. Churchland, *Scientific realism and the plasticity of mind* (Cambridge: Cambridge University Press, 1979), p. 2. The following references to Churchland are to this work. Also, see his *Matter and Consciousness* (Cambridge: MIT Press, 1984), ch. 3.
33. Cf. Saul Kripke, *Naming and Necessity* (Cambridge: Harvard University Press, 1972), pp. 144–145.

34. It is ironic that in Russia, where the official ideology is dialectical materialism, greater credence and more financial support is given to psychical research than in the United States. See Sheila Ostrander and Lynn Schroeder, *Psychic Discoveries Behind The Iron Curtain* (Englewood Cliffs, N.J.: Prentice-Hall, 1970).

35. John R. Searles, *Intentionality* (Cambridge: Cambridge University Press, 1983), p. 262. The following references to Searle are to this work.

36. Cf. Richard H. Schlagel, "The Mind-Brain Identity Impasse," *The American Philosophical Quarterly,* Vol. 14, July 1977), pp. 231–237.

37. For an interesting attempt to establish a new scientific paradigm, a holographic model, for interpreting the mind-body problem, see Ken Wilber (ed.), *The Holographic Paradigm* (Boulder: Shambhalla Publications, 1982); also David Bohm, *Wholeness and the Implicate Order* (London: Routledge & Kegan Paul, 1980), ch. 6.

38. Cf. Richard Rorty, *Philosophy and the Mirror of Nature* (Princeton: Princeton University Press, 1979). The following references to Rorty are to this work.

39. Cf. Introduction, f.n. 11.

CHAPTER II

THE ORIGINS AND BASIS OF KNOWLEDGE

IN the previous chapter I argued that the traditional problems of knowledge were outgrowths of the changing world view and empirical investigations of the seventeenth century. Along with the brilliant scientific attainments of "the age of genius," as Whitehead called it, various paradigms of philosophical inquiry took hold with such tenacity that they are only now being seriously challenged— just as the Newtonian world–machine paradigm lasted until the beginning of the twentieth century. Far from inadvertently misusing ordinary language, the founders of modern philosophy intentionally devised unusual linguistic expressions and uses to articulate the new questions and possible solutions raised by the prevalent intellectual revolution—just as Freud and Einstein introduced novel terms to express their psychoanalytical and relativity theories: for example, "*unconscious* desires," "*repressed* ideas," "*curved* space," and "*dilation* of time." In terms of previous usage and conventions, these expressions would have been meaningless, contradictory, or at least highly paradoxical.

Rather than ascribe such problems and the various theoretical strategies devised for dealing with them to "category mistakes" or "linguistic muddles," they can more accurately be attributed to the utilization of certain empirical models in explanation (such as the camera obscura) and to theoretical presuppositions (such as the prevalent belief that the nervous system operated hydraulically). Just as it is hardly a "category mistake" today to interpret the central nervous system as functioning analogously to an electronic computer, the model fitting at least the empirical discovery that neurons conduct coded electrical discharges, so our philosophical forebears attempted to explain how the nervous system functioned in terms of the knowledge available to them.

Moreover, it was pointed out that the scientific discoveries of the present century, which are at least as revolutionary as those of the seventeenth, have produced new approaches to problems requiring revisions of the older epistemic positions. In particular, the radical Cartesian paradigm which claimed that knowledge of our mind and its contents had an epistemic certitude, and thus an ontological priority over the body, was not born out by the theory of evolution or by developments in neurophysiology. While Descartes could think of the human mind as a "distinct" spiritual substance, and human capabilities as being completely discontinuous with what he believed to be the purely mechanical behavior of animals, it is difficult to maintain these positions today in light of recent investigations in comparative vertebrate anatomy, genetics, neurophysiology, and animal behavior.

Not only has it been found that apes can be taught to use abstract symbols and to communicate by sign language–disproving the idea of a complete discontinuity between the mental functioning of human beings and of our simian progenitors—but, conversely, as the recovery of feral children indicates, there is no humanoid essence that manifests itself naturally if the child is isolated from contact with other persons. In addition, comparative anatomical studies show a continuous evolution in morphological systems and their correlative functions, with "ontology recapitulating phylogeny," while the remarkable discoveries in genetics have established that the maturation of all living organisms is dependent upon a nearly identical DNA code.

Accordingly, we shall begin our inquiry into the origins of knowledge not from the older Cartesian paradigm, but from the newer perspective of an evolved human organism possessing a central nervous system whose main function is to process information derived from the physical world in order to respond appropriately and creatively to this external environment. However, as we are concerned with knowledge rather than with behavioral responses, such knowledge being a function of higher cerebral processes that we know relatively little about, we shall concentrate on the conscious derivation of this knowledge ignoring, for the present, whatever neural wiring or circuitry underlies it. Even here, however, empirical research in both neurophysiology and cognitive psychology offers information directly relevant to such traditional philosophical problems as the empiricist versus rationalist (or nature versus nurture) controversy.

For example, philosophers are still disposed to quote William James' Humean characterization of an infant's original experience as a "blooming, buzzing confusion," though recent psychological investigations have thoroughly disproved it. As James depicted it, an infant's initial experience would consist of a chaotic, kaleidoscopic mix of various sense qualities totally lacking in form and structure and completely random in sequence, following the British empiricists' analysis of the original contents of consciousness into discrete, irreducible, sensory ideas or impressions. Thus patches of color, tactual and olfactory sensations, and discrete sounds and smells were thought to float before the mind of a child in a completely haphazard, disconnected way. How they subsequently became synthesized into our precepts or ideas of a seemingly independent world of physical objects and causal situations was something of a mystery.

Hume, literally basing his model of interpretation on the atomic theory, claimed that these original sensory elements organized themselves into images owing to the "gentle force of association" emanating from them: in other words, forces of attraction generated by the resemblance, contiguity in space and time, and causal sequence of the sense elements themselves. Thus those impressions which occurred proximally in space and time, or which resembled each other, or which followed in succession would mechanically associate under the influence of such attracting forces. Unlike other philosophers of the period, Hume did not attribute any active role to the mind, defining it simply as a "bundle of sensations."

This passive notion of the origin of percepts was completely rejected by Kant who emphasized the active role of the mind in synthesizing the "sensory manifold" with inherent cognitive forms or structures, and in constructing images and ideas with the schemata of the imagination. While recent research indicates that the process is more complex than either the empiricist or Kantian account, it favors Kant's interpretation that percepts and knowledge are dependent upon complex synthesizing processes based on innate structures—although now these innate structures and underlying integrative processes would be located in the brain, rather than attributed to an independent mind (or "transcendental ego"). Thus, in contrast to the empiricists who claimed that our original experience consisted of discrete sensory elements devoid of any initial form or structural organization, and that all perceptual information (such as visual depth perception and object constancy) had to be "learned from experience," recent investigations of perceptual processes and of the visual world of infants disprove these contentions.

For example, experiments on newly born chicks demonstrate that they can immediately discriminate the shapes of three-dimensional objects,[1] and that they possess binocular depth perception.[2] Moreover, the experiments were ingeniously carried out to determine whether these abilities, although immediately displayed, might not depend upon some subtle learning rather than on an innate capacity, but the results convincingly showed that these behaviors were unlearned, produced by an inherent ability to integrate visual information based on the chick's genetically programmed visual system. These results should not be surprising since an innate proclivity to discriminate and peck at grains of corn enhances a chick's survival.

The evidence for innate depth perception was reinforced by another experiment, "the visual cliff," in which it was shown that newborn animals and infants at the creeping stage refused to move onto a sheet of glass exposing a visual depth (or "cliff"), although they could *feel* the solid glass. When called, infants refused to crawl onto the depth–revealing glass surface despite the fact that by patting it they could feel it was hard, while a one–day-old goat, when placed on the glass sheet, immediately froze, assuming a seated defensive posture with its hind legs limp and its front legs stiff.[3]

The fact that the infant as well as the animals tested would not move onto the depth–revealing surface, even though information about its stability or firmness was transmitted at the same time, indicates the priority of visual over tactual data for depth perception. This, in turn, tends to invalidate Berkeley's empiricist argument that binocular vision is completely learned, and that the sense of touch informs vision of the distance, size, shape, and solidity of objects. Moreover, when adults were tested as to whether tactual or visual data was dominant in determining distance and size, it was found that vision was not only paramount, but so dominant that when a conflict arose in the information supplied by visual and touch receptors, most subjects were unaware of the discrepancy, having accepted the overriding evidence of their vision.[4]

Other investigations have determined that infants can perceive form without having to learn to do so, in contrast to the empiricist's view. According to the experiments of Robert L. Fantz and T. G. R. Bower, primate infants have an innate capacity to perceive not just light, color, and movement, but shapes, patterns, and sizes. A five-month-old chimpanzee indicated a visual preference for certain objects, while human infants from one to fifteen weeks of age evinced an unlearned, inborn ability to discriminate patterns, preferring a painted face to a bulls-eye, checkered, or scrambled pattern. Moreover, the child's capacity for depth perception, form completeness, size and shape constancy, along with orientation discriminations is much greater than could be accounted for on purely empiricist principles. As Fantz summarizes these findings, emphasizing that further research will bring refinements in the interpretation:

> ... the results to date do require the rejection of the view that the newborn infant or animal must start from scratch to learn to see and to organize patterned stimulation. Lowly chicks as well as lofty primates perceive and respond to form without experience if given the opportunity at the appropriate stage of development. *Innate knowledge of the environment* is demonstrated by the preference of newly hatched chicks for forms likely to be edible and by the interest of young infants in the kinds of form that will later aid in object recognition, social responsiveness and spatial orientation. *This primitive knowledge* provides a foundation for the vast accumulation of knowledge through experience.[5] (Italics added.)

In their conviction that perception of forms and distances had to be learned from experience, the empiricists were misled (as Descartes and Kepler had been) by the model of the camera obscura. Believing that what one perceives are static two-dimensional images projected onto the retina, they thought that all perception of distance, solidity, size, and shape constancy had to be learned from cues supplied by other sense modalities, such as touch and the kinesthetic sense of ocular convergence when looking at objects of varying distances. Thus the infant had to learn from experience (to infer from the limited information supplied by the retinal images supplemented by other sensory clues) to be able to perceive sensory patterned, stable, solid objects arrayed in a three-dimensional space, as adults do. Again, however, experimental evidence does not support this interpretation.

Given an innate visual system, as well as an inherent cerebral capacity for processing data, it is now believed that instead of responding to a procession of static images, the brain of the child processes a variety of constantly fluctuating visual data consisting of linear patterns, textures, densities, and orientations, along with kinaesthetic sensations and information provided by motion and binocular parallax (motion parallax designating the fact that when one moves one's head, closer objects are displaced more markedly than distant ones, while binocular parallax refers to the changing angles of convergence of the two eyes when focusing on objects at various distances). Thus, instead of the infant's mind having to construct in a piecemeal way a picture of the world by successively adding independent elements, relations (spatial, temporal, and causal), and structure on the basis of accumulated experience, it is the central nervous system that has been genetically programmed to be capable of processing the relevant information into perceptions of patterned, stable, three-dimensional forms.

Rather than being limited to perceiving the restricted features of the retinal image, the child's visual system has been programmed to respond to the objectively real characteristics (within the context of experience) of the ordinary macroscopic world.[6] This is not to claim, however, that the child perceives the same world as the adult. As compared to adults, the child's processing capabilities are very limited, probably assigning priority to information that has particular survival value. Moreover, the brain weight of the infant is only about one-quarter that of the adult, and though the total number of neurons is practically formed by birth, the increase in glial cells, the growth of mylinated axons, and the spreading of interconnecting dendritic and axonal pathways occur postnatally as a result of further physiological experience and development.[7] So though the child does not begin experience on a completely passive and unstructured basis—the proverbial "blank tablet"—most of the connectedness of its neuronal network, the underlying physical basis of his experience, is established after birth.

Essential to this development is an aspect completely ignored by classical epistemologists, although it has been appropriately emphasized by contemporary empiricists, such as the pragmatists—the indispensable element of bodily movement and exploration in the acquisition of knowledge. For Kant, as well as for the British Empiricists and Continental Rationalists, the development of knowledge was not seen as dependent upon any exploratory movements of the body. From Descartes' solitary meditations by the warmth of a fire, to Kant's redefinition of metaphysics as the critical investigation of the *a priori* or transcendental presuppositions of all knowledge, no consideration was given to the significance of active bodily interaction with the world as necessary for the growth of knowledge. In place of the notion of a human being as a curious organism actively exploring the world (which is true not only of infants and animals, but of all living organisms), the classic model was that of an immobile reflective mind (a kind of television or computer screen) on which sensory inputs from the external sense organs (a system of static sensors) were relayed, transformed, and recorded. Descartes did refer to the mind as a kind of pilot directing the body, but his paradigm of analysis in the *Meditations* was that of a secluded mind reflecting on its own contents.

Here again, recent empirical investigations show this inactivist presupposition to be mistaken. For example, in an experiment by R. Held and A. Hein which tested the older empiricist thesis that knowledge could be acquired simply from an original array of sensory elements, independent of any bodily movements, the results were definitely negative. The experiment consisted of suspending two kittens from a carousel in such a way that while both received the same visual data, only one kitten received these data as a result of its own physical exploration of the environment. That is, the active kitten was attached to the carousel in a manner that permitted it to walk about freely, but the other kitten was suspended in a way that prevented it from initiating any movements, although because it was swung around by the activity of the other kitten, it received the same visual stimuli.[8] The result was that while the mobile kitten had synthesized its visual data with its kinesthetic sensations so that it could behave normally when released from the carousel, the passive kitten had learned nothing; it was incapable of reacting to the visual cues in the environment when freed from the apparatus. This demonstrates that bodily movements are necessary to establish the kinds of integrated neuronal connections that underlie normal awareness and appropriate perceptual responses to the world—a conclusion reinforced by the discovery of children severely retarded as a result of being brought up in closets or otherwise isolated from any contact with people.

Jean Piaget, the developmental ("genetic") child psychologist, has particularly emphasized the continuity between what he calls our "sensorimotor intelligence" (sensory and physiological responses to the environment) and higher intellectual functions, suggesting that there might even exist "a functional nucleus of the intellectual organization which comes from the biological organization in its most general aspect," such that "this invariant will orient the whole of the successive structures which the mind will then work out in its contact with reality."[9] Without going into the details of Piaget's elaborate system, one can say that he interprets the child's cognitive development as based on earlier modes of exploratory and adaptive behavior by constructing various "schemata" for coping with the world. These schemata are successively developed mental structures into which incoming sensory data are made to fit, analogous to Kant, except that Kant's mental structures were innate and invariant.

As in biological growth, certain *invariant functions,* such as "organization" and "adaptation," underlie *transitory structures* (such as logical and mathematical operations, spatial and temporal relations, concepts of causality and substance) which the child develops in its interactions with the environment. Thus while the *functions* are permanent and invariant, the *structures* change systematically as the child develops. Additional "functional invariants," characteristic of all biological systems, are "assimilation" and "accommodation." Assimilation refers to the organism's ability to take in and utilize something from the environment (such as food and sensory stimuli), while accommodation designates the organism's ability to mediate what is being assimilated by modifying or correcting its response to it. Since the organism is constantly active, there is a continuous interaction of these processes which, in the ideal situation, results in "equilibrium." But this equilibrium is always temporary and in tension so that the individual is continuously modifying its structures, a process he calls "equilibration." These are the mechanisms Piaget believes control cognitive growth so that one can discern definite developmental units called "periods," "subperiods," and "stages" through which the child must pass if he or she is to develop. More precisely delineated, these units are "the sensorimotor period" (with six stages, including the stage of preoperational thought), "the concrete operations period" (with two subperiods), and "the formal operations period." The growth of knowledge thus depends upon a continuous interaction between the child and the world with increasingly more sophisticated cognitive structures being elaborated as it passes through necessary stages of development. This thesis, that a proper understanding of knowledge must take into account the active interaction of the knowing organism with the environment, in contrast to the classical model of an immobile spectator confined to the inner recesses of the brain, has also been a central thesis of pragmatism.

Another element essential to cognitive development overlooked by the classical empiricists and rationalists is the role of other human beings in the education of the child. Granted that the infant is not just a malleable sentient organism, but has inherited a complex, genetically programmed central nervous system that predisposes it to respond and develop in certain ways and stages, this development is not mechanical, automatic, or self-contained. In addition to sensations derived from the stimulation of the sense organs, the child has drives, feelings, and emotions that play a crucial part in its normal growth. As D.W. Hamlyn has emphasized, cognitive development in the child is not "simply. . .the product of an interaction between the individual's innate structure and the structuring influences to which he is subject because of the expression of his innate tendencies,"[10] but also depends upon affection, love, training, correction, and such. The inherited tendencies or dispositions of the child are guided and molded by the continuous influence, encouragement, and correction of the people caring for it. These are the "forms of life" that Wittgenstein considered so important as a contextual background for the learning and the use of language, but which are equally important in nurturing the child's conception of truth and acquisition of knowledge.

Since the infant is not born with innate knowledge as such, anymore than it is born possessing a particular natural language, but with the necessary structural and functional prerequisites for learning, the precise forms of knowledge learned by the child depend upon information taught it by other people. Feral children brought up by animals manifest forms of behavior and knowledge more animal-like than humanoid. As Hamlyn maintains, the child is not just a computer, but a person nurtured within a cultural environment composed of other human beings (cf. p. 102).

Although recent investigations do not support the empiricist's conception of knowledge, they do not imply a simple nativist interpretation either. The origin and development of experience and knowledge cannot be explained simply. While in contrast to the empiricists we must acknowledge the role of certain *innate structures and capacities,* these are not necessarily developed and operative at the time of birth, but require bodily movement, physiological maturation, external stimuli, and a normal human setting for their realization. Thus, in opposition to Kant's transcendental philosophy, the construction of knowledge cannot be attributed simply to the function of preformed *a priori* structures alone.

Instead of either empirical conditioning or inherent *a priori* forms, experience and knowledge are based on the complex interrelationship of (1) innate hereditary structures and functional capacities, (2) exploratory movements generating and integrating sensory stimuli, (3) biological maturation, and (4) a culturally reinforcing process of learning. Thus normal cognitive development depends upon the activation of our inherited capacities by appropriate environmental stimuli during optimal maturation periods. If any of these components are omitted or brought into play at the wrong time, learning or development becomes distorted, retarded, or even blocked altogether. For example, there is an optimal time for learning a language so that if one attempts to teach a child a language after this period, as has happened with feral children, a great deal of effort is required and the final "language competence" is very limited.

Although the discussion so far, including the experimental evidence, has relevance for the traditional empiricist–rationalist controversy, one might question whether it has significance for a contemporary understanding of knowledge. The answer is definitely affirmative if one is concerned, as were traditional philosophers, not with questions of how terms such as 'belief,' 'knowledge,' and 'truth' are used in ordinary language, but with describing the origins, development, nature, and limits of knowledge. Such inquiries are especially pertinent to the problem of what must be presupposed for knowledge to be possible (Kant's question), and to understanding how we acquire the various forms of knowledge we do. For example, since the principle function of our sensory receptor system is to discriminate among various physical stimuli so as to convey to the organism necessary information about the external world in the form of sensory qualities, shapes, positions, and movements, these data necessarily become the original basis of knowledge. And since the nervous system itself functions in such a way as to discriminate, abstract, and then integrate information into reconstructed perceptions, it is not surprising that the *conscious* creation of knowledge, as we shall find, occurs in a similar manner.

Moreover, recognizing the underlying role of the central nervous system in making possible certain conscious activities can remove unnecessary philosophical perplexities. For example, in the *Philosophical Investigations,* Wittgenstein was puzzled as to how we can discriminate colors and shapes, asking

> what does 'pointing to the shape,' 'pointing to the colour' consist in? Point to a piece of paper.—And now point to its shape—now to its colour...How did you do it?—You will say that you 'meant' a different thing each time you pointed. And if I ask how that is done, you will say you concentrated your attention on the colour, the shape, etc. But I ask you again: how is *that* done?[11]

After expressing this apparent bewilderment as to how discrimination of forms and colors occurs, Wittgenstein goes on to conclude:

> ... because we cannot specify any *one* bodily action which we call pointing to the shape (as opposed, for example to the colour) we say that a *spiritual* [mental, intellectual] activity corresponds to these words (brackets in the original).
> Where our language suggests a body and there is none: there, we should like to say, is a *spirit* (Part I, sec. 36).

Wittgenstein was objecting to the introduction of spirit (or consciousness) to explain the essential cognitive functions of discrimination and intention. But in fact, discriminating is *both* a bodily (or neurological) and a spiritual (or conscious) process: we can *consciously* distinguish sensory qualities and forms because our sensory neurological systems respond to different physical stimuli in such a way as to produce qualitative discriminations on the conscious level (though we are not aware that this is occurring). We can describe either aspect of the process, the neurological or the conscious, but discriminating and designating are essentially cerebral–conscious capacities. Thus Wittgenstein's question *how* it can be done is not mysterious in the way he would make it appear. *Too often analytical philosophers try to provide a conceptual explanation for a capacity that can only be explained by empirical investigations.*

If the child (or animal) were unable to do the following: (1) discriminate among sensory qualities and forms, (2) focus it's attention on them to fix them in it's visual field, (3) recognize similarities and differences among the discriminated data, and (4) retain some nervous imprint (encoding) resulting in conscious memory or recognition when similar data recur, then it would be difficult to understand how any knowledge could be possible. But from their first movements chicks are capable of binocular discrimination of geometrical forms, while as early as one month a child can select among variously colored patterns and evince recognition of the reappearance of a human face, a feat essential for the reinforcing schedule. Moreover, animals and humans can distinguish static from moving objects, usually attending to the latter, while fragmentary or incomplete stimuli can arouse the same response as full stimuli. Finally, as the organism matures it retains some neuronal storage of past experiences which enables it to recognize recurring stimuli and anticipate absent phenomena either on the basis of partial stimuli (as when a dog becomes restive on hearing it's owner's footsteps) or no stimuli (as when a child cries for attention).

This recognition of, or anticipatory response to, recurrent stimuli is probably the basis of one of the crucial elements of knowledge, *concept formation*. For just as discrimination among sensory qualities and forms is essential for any meaningful response to the environment, so the formation of concepts is essential for any knowledge. By a concept I mean the conscious *recognition* or *identification* of occurrent stimuli, as well as the *anticipation* of familiar absent stimuli. As Hamlyn states, in the book previously cited, "to have the concept of X is to know what it is for something to be an X. . . ." (p. 74). Just as successive neuronal imprints of the same object would seem to result in a *conscious sense* of recognition or identification once the imprinting process has occurred, partial stimuli, or even internal physiological stimuli (as when a child is hungry), would seem to arouse in the child a *conscious anticipation* of an entity. Then, as the neuronal reinforcement continues, the child is not only conditioned to respond behaviorally to external stimuli, it is capable of forming *internal schema* of perceived stimuli.

Although there would seem to be no absolute distinction, as development occurs, between meaningful behavioral responses and rudimentary internalized schema, these latter are what are particularly meant by concepts. In so far as animals are capable of recognitional or anticipatory behavior, one could say that they possess *implicit* concepts; but it is not until a child begins to form internalized schema that it can be said to possess *explicit* concepts. That these internalized concepts are schema rather than specific images seems to be indicated by such evidence as an infant's acknowledgment of any woman as its mother, or identification of similar canines as dogs.

Thus prior to the learning of language a child forms concepts or internal schema. Moreover, the classic example of the ape finally learning to use a stick to reach a banana might, to some extent, depend upon 'seeing' possible connections among internal schema. But even if this is not the case, the fact that chimpanzees who had been trained to use symbols to respond to commands, answer questions, and convey information *spontaneously generalized* the use of symbols, transferring the application of signs from objects to pictures and replicas, and subsequently began using signs or symbols in novel ways,[12] seems to indicate at least an implicit use of concepts. For the child, these rudimentary concepts later become the meaning of words. Just as apes learn to use symbols and signs in place of objects in *sign* communication, so children learn to use meanings and words in place of objects in *linguistic* communication. Thus I agree with Piaget and Hamlyn that the development and use of concepts precedes the learning of a language.[13]

When Helen Keller rediscovered that sensations, objects, and occurrences (such as the flow of cool water over her hand) have names and meanings, a whole new cognitive realm opened up for her. Though she was prevented from conversing directly with nature through vision and sounds, the realization that felt experiences have names, and thus could be identified, understood, and referred to as distinct phenomena, seems to have been a kind of revelation akin to sight and hearing. From the moment she knew she could communicate her reactions, feelings, and thoughts to others by sign language, and in turn, receive theirs, she became aware of the unlimited possibilities of enlarging her experience. For it is language that frees us from the captivity of the here and now, enabling us to think about the future and the past, the actual and the counterfactual—to envision possible worlds.

Because psychological and philosophical behaviorists are reluctant to acknowledge inner conscious states or processes (such as images, meanings, and imagining), they often bypass the role of concepts in thought, despite the fact that recognizing the function of concepts is indispensable for any adequate theory of knowledge. Although much of our behavioral response to the world is habitual or automatic, so that usually we are not explicitly aware of using concepts, as soon as we begin to think their role becomes obvious. For example, concepts are manifested in such ordinary experiences as identifying phenomena, interpreting signs, following instructions, drawing figures, using tools, calculating ("in one's head"), and talking and reading.[14] (To appreciate the importance of concepts, recall the last time you encountered an unknown word or an unfamiliar object—one for which you had no identifying concept— and your subsequent inability to respond intelligently.)

While it is true, as Ryle insisted, that we often use pencil and paper to work out mathematical and other theoretical problems, surely the thought process is not just manipulating a pencil. Moreover, many people have had the experience of reaching "a block" in trying to solve a problem, and then finding that the "answer comes to them later" when they are not even working on the problem (as in the classic case of Kukulé seeing in the dream image of a snake biting its tail the solution to the hexagonal molecular structure of the benzene ring). Again, the cerebral aspect of our cerebral–conscious processes is undoubtedly at work when we "see" the solution to the problem.

When we reflect on what occurs as we learn a new principle or theory, the subtle interrelation of words, images, and concepts can hardly be denied. If, for example, a physicist begins a lecture by saying, "I am going to discuss Heisenberg's principle of indeterminacy," the words could be intelligible to the audience (if it has already heard of Heisenberg and his principle) without their having any real meaning. But the comprehension of the audience will increase if the lecturer continues by stating: "while the size of macroscopic objects is such that their measurement does not disturb them, when one considers measuring the position and momentum of a subatomic particle such as an electron (realizing that in order to measure it there must be an interaction between the particle and some detecting instrument, such as a gamma wave microscope), the precision of the measurement does become problematical because of the interaction of the wave with the particle. That is, the interaction introduces an element of "indeterminacy" in the attempt to measure, simultaneously, both the position and the momentum of the particle."

As I am using the term, it would be impossible to understand such a lecture without acknowledging the role of concepts such as 'electron,' 'subatomic particle,' 'momentum,' and 'gamma wave.' Not that these concepts must explicitly come before the mind as one hears the lecture, but if the words had no meaning, the lecture would be incomprehensible. As long as discourse is intelligible, it is *as if* only the spoken and heard words were involved. In the graphic image of Bosanquet, "Language is so transparent that it disappears so to speak, into its own meaning, and we are left with no characteristic medium at all."[15]

Moreover, if we only had to take into account the flow of words, but not their meanings, in communication, then learning a foreign language would pose no problem once we acquired the capacity to discriminate the phonetic structure. Though usually we hear and speak in such an effortless way that it is possible to overlook the essential role of meanings or concepts in communication (as I am using the terms, meanings are what concepts become when they are associated with words), whenever we cannot remember a name or cannot find *le mot juste* to convey a thought, or when we realize that what we have said is unclear or even contrary to our intention, we are made aware of the essential role of concepts or meanings as we speak, though normally they occur subcortically or subconsciously. These implicit meanings constitute thought and enable us to read, listen, and speak intelligently. As I pointed out in the last chapter, one of the effects of the injury to the left parieto–occipital area of Zasetsky's brain was that he had difficulty in attributing meanings to the most familiar words. In addition, there are cases of individuals who have suffered strokes in the Broca's or Wernicke's area of the brain so that either their *verbal* ability to express thoughts has been severely impaired, or they have lost the capacity of making *sense* of what they are saying and making it *appropriate to* the situation (one of Chomsky's three criteria of human speech), even though their ability to form grammatically correct sentences has not been adversely affected. As described by Howard Gardner:

According to classical aphasiological theory, Broca's area is responsible for the conversion of ideas, perceptions, and intended messages into smoothly articulated patterns of speech, structured by the appropriate syntactical forms. Lesions there will produce the paucity of speech found in Mr. Ford ["impossibility of understanding just what he was talking about"]. In the other major type of aphasia [Wernicke's], Broca's area is still intact, but appears to be largely dissociated from direction by the patient's conscious ideas and intentions. Present is the outward *form* of normal speech, its flow, its connective and adjunctive words interspersed among the more crucial substantives; largely absent is direct and accurate reference to specific elements in the environment, as well as the capacity for coherently interweaving particles and substantives into a meaningful whole. Wernicke's area appears crucial in two functions: relating incoming sounds to the representations (or "meanings") which allow understanding of discourse; selecting and arranging meaningful units for eventual conversion into comprehensible, coherent speech. These functions are obviously essential for understanding and emitting language.[16] (Brackets added.)

Similarly, either a person's ability to read aloud can be affected or their capacity to understand what they are reading. In my opinion, this empirical evidence strongly supports the necessity of distinguishing between the use of words in speaking and their understood meanings.

That meanings are indispensable for explaining normal comprehension when discoursing or reading may seem so obvious that one might wonder why I belabor the point. The reason is that a number of analytic philosophers with a disposition towards behaviorism, such as Wittgenstein, Ryle, and Quine, despite what appears to be definite empirical evidence to the contrary, have denied that reference to mental contents, such as meanings, contributes in any way to our understanding of the acquisition and use of language. Although we shall further pursue the question of meanings in the following chapter, a preliminary discussion is appropriate at this point, since it is crucial to any theory of knowledge.

Wittgenstein, for example, not only denied that thinking was an "inner process" (in his attack on "inner experiences" and "private languages"), he also denounced the role of meanings in his interpretation of how we learn and use language. As regards meanings, he said: "When I think in language, there aren't 'meanings' going through my mind in addition to the verbal expressions: the language is itself the vehicle of thought."[17] But if thinking involves only verbal expressions, then so do talking and listening, so that if "language is itself the vehicle of thought," then when I converse "there aren't meanings going through my mind," only the sound itself. According to this account, instead of the common expression "what do you mean by that?," we should have the expression "what do you say by that?," which would make little sense. And though Ryle did not directly address the problem of meanings in *The*

Concept of Mind, he eschewed any reference to mental contents or processes, asserting that the "phrase 'in the mind' can and should always be dispensed with. Its use habituates its employers to the view that minds are queer 'places', the occupants of which are special-status phantasms."[18] There is, of course, an element of truth in this, yet when we try to make our ideas more precise, as I am attempting to do now in writing this, it does not seem that they are mere "phantasms."

Though his reasons are somewhat different, Quine also derides the concept of meaning, declaring that "uncritical semantics is the myth of a museum in which the exhibits are meanings and the words are labels."[19] However, unless one conceives of meanings as Lockean ideas "that stand before the mind when a man thinks," thereby replacing objects as the referents of our words, Quine's criticism seems misplaced. On the conscious level, meanings are what enable us to understand words and guide their use in discourse, whatever their neurological status might be. Because certain lesions result in a pathological disassociation of meanings and words, brain damaged patients such as Zasetsky could not identify referents of such common words as 'nose,' and 'eyes,' though the terms had a nagging ring of familiarity. This happens also in daily life when we cannot recall the meaning of a term though we realize we should know it. Correcting Quine's museum metaphor, the exhibits are objects in the world (not meanings), while the labels are words which, if they are intelligible (either in the museum or in everyday discourse), must be meaningful to the individual. Yet Quine goes on to state:

> Semantics is vitiated by a pernicious mentalism as long as we regard a man's semantics as somehow determinate in his mind beyond what might be implicit in his dispositions to overt behavior. It is the very facts about meaning, not the entities meant, that must be construed in terms of behavior (p. 27).

Rather than meanings being "implicit" in our *cognitive* recognitions, recollections, and reflections, Quine limits them to what is "implicit" in our *behavior.* Thus he agrees with Wittgenstein's view that "language is itself the vehicle of thought," not meanings or concepts. But how many of us actually think in specific words, rather than with implicit meanings, images, or concepts? I rather agree with Henry Mehlberg who says "the occurrence of talking in one's head during the process of problem solving or of performing this activity in some specific language. . .is certainly exceptional."[20] Quine would like to exorcise meanings because he believes the only evidence available to account for language learning and use is behavioral, therefore reference to meanings does not explain anything, despite the fact that the term is used so frequently in explicating such common experiences as mutual understanding or misunderstanding, and translating from one language to another. However, when we talk of knowing the meaning of an expression or knowing that two statements express the same meaning or are equivalent in meaning, Quine maintains that we should not delude ourselves into thinking there are meanings in addition to the verbal expressions. His behavioral criterion of knowing the meaning or the sameness of meaning of expressions is *the ability to substitute one verbal expression for another.* As so often is true in Quine's writings, these crucial positions depend upon single examples, in this case, the example of his two young children. When asked if he knew what 'Eighty-two' meant, his small son replied "No", but when asked what 'Ottantadue' meant, his daughter answered "Eighty-two".[21] According to Quine, the difference in their replies illustrates the fact that while his son could not give a substitute expression, his daughter could, demonstrating that understanding or explicating terms depends upon substituting equivalent expressions, not giving meanings. As he says, instead of talking of "knowing the meaning, and of giving the meaning, and of sameness of meaning," we should talk of "understanding an expression, or talk of the equivalence of expressions and the paraphrasing of expressions" (p. 86).

This, however, does not explain what "understanding" an expression consists of. In order to avoid a "pernicious mentalism" he advocates accepting the "understanding" and "equivalence" of expressions as unanalyzable behavioral dispositions, while I think they can be further explicated by meanings and concepts. These meanings are what constitute understanding and equivalence when we search for the right meaning of a word or the correct translation of an expression. This is supported, it seems to me, by the fact that we know whether we understand an expression *prior* to our manifesting this understanding in verbal behavior. When listening to a conversation in a foreign language we usually know, apart from any behavioral response, whether we understand what is being said, though of course sometimes we are mistaken and manifest this in an incorrect or inappropriate verbal retort. Thus I believe that there is an essential, irreducible *mental* aspect involved in the use of language, although its "perniciousness" can be avoided if we admit that mental states are dependent upon (or located in) cerebral processes.

Furthermore, one can think of counter examples to the one provided by Quine. Often, when we are asked to give the meaning of a term or explain something that we have said, we cannot immediately find an equivalent expression, as Quine demands, yet by reflection we attempt to explicate the word or statement. In these cases, we have a *sense* of what we want to say before we can find the exact words to express it. What guides us in the search for the right expression, and what indicates that we have found it? In my view, this is precisely the role of meanings or concepts as implicit accompaniments of our thought. When we listen to a lecture or a discussion, we *never* translate what is being said into other words, unless we are concerned with recounting the lecture specifically to someone else; instead, we simply understand or do not understand the intended meaning of the speaker. Similarly, when learning a foreign language, we initially translate foreign expressions into our native language, just as when we begin speaking a foreign laguage we first translate from our own language into the foreign one. Later, when we gain proficiency in the new language, we understand the expressions immediately because we have learned their meaning without having to match them with expressions in our own language. The point is, we normally do not think in words but with implicit meanings, images, and schemata. Furthermore, since Wittgenstein and Quine deny that there are meanings in addition to words, their position would imply that a child is incapable of thinking before learning a language, a thesis denied by Piaget.

I do not believe, therefore, that language is acquired "on the evidence *solely* of other people's overt behavior under publicly recognized circumstances,"[22] as Quine maintains, but also as a result of reflecting on the meanings of what we have read and discussed. Quite often we resort to visualizing and imagining when trying to grasp the sense of what is meant. Since we cannot explain this process neurologically, as Quine recognizes, I think we should utilize whatever explanatory data are available, instrospective or cognitive, as well as behavioral. Even Quine finds it difficult to dispense with terms such as 'meaning,' 'understanding,' and 'thinking,' although he attributes this to cultural or linguistic inertia rather than to any experiential necessity. But do we really know?

To illusrate further the arbitrariness of a purely behavioristic account of language, consider Einstein, who partially attributed his success in conceiving the special theory of relativity (which began with his reflections on the velocity of light when he was sixteen years old) to his ability to think by means of concrete examples (*gedankenexperiments*), rather than in words (Sheldon Glashow also has said that he is a visualizer). According to his sister Maja, Einstein was late in learning to talk (causing concern in his parents whether he might be somewhat retarded), which resulted in an unusual capacity to think in images and concepts, instead of words.[23] In addition, there is the anecdote reported by Philipp Frank, that arranging to meet on a bridge, Frank was concerned to specify the time so as not to keep Einstein waiting, while Einstein responded that it really didn't make any difference if he were kept waiting, since he could think about his problems on a bridge as well as any other place. This, it seems to me, belies Wittgenstein's, Ryle's, and Quine's view, that thinking is not an "inner mental process" but a behavioral response of some kind.

In his attack on "mental processes," Wittgenstein asserted, "Try not to think of understanding as a 'mental process' at all."[24] But if understanding is not a "mental process," what is it? Surely it is not merely the behavioral manifestation of indicating that one has arrived at the solution to a question. If this were true, Archimedes' purported exclamation, "Eureka," after realizing he had discovered the law of specific gravity would be equivalent to the discovery itself! How would Wittgenstein, Ryle, and Quine explain the musical capacities of composers like Mozart who could "create an entire overture 'in their heads' before even setting pen to paper"?[25] While some people write with their pens, typewriters, or computers, others carefully "think out" what they want to say before recording it. Furthermore, apparently Wittgenstein, Ryle, and Quine were never kept awake at night compulsively mulling over a problem. These considerations illustrate again the inherent weakness of philosophical analyses as a methodology for recommending conceptual or linguistic revisions, if undertaken independently of a careful review of the relevant empirical evidence. As presented especially by Wittgenstein and Ryle, such analyses have a distinctive *a priori* character, in opposition to more empirical inquiries.

Thus far it has been maintained that concepts or meanings are essential for thought and knowledge, and that they probably originate as conscious recognitions and anticipations correlated with behavioral adaptations. As the contents of conscious states they can range from eidetic and concrete images in one extreme, to abstract general schema (such as the concept of 'a triangle' or 'a man,' the meanings of which troubled both Berkeley and Russell) and non-visualizable concepts (such as 'infinity,' 'the four-dimensional manifold of events,' and 'things in themselves'), on the other. Unlike the philosophers just discussed, Kant went to considerable effort in "The Schematism Of The Pure Concepts Of Understanding" to explain how this is possible.

> Indeed it is schemata, not images of objects, which underlie our pure sensible concepts. No image could ever be adequate to the concept of a triangle in general. It would never attain that universality of the concept which renders it valid of all triangles, whether right-angled, obtuse-angled, or acute-angled; it would always be limited to a part only of this sphere. The schema of the triangle can exist nowhere but in thought. It is a rule of synthesis of the imagination, in respect to pure figures in space. Still less is an object of experience or its image ever adequate to the empirical concept; for this latter always stands in immediate relation to the schema of imagination, as a rule for the determination of our intuition, in accordance with some specific universal concept. The concept 'dog' signifies a rule according to which my imagination can delineate the figure of a four-footed animal in a general manner, without limitation to any single determinate figure such as experience, or any possible image that I can represent *in concreto,* actually presents. This schematism of our understanding, in its application to appearances and their mere form, is *an art concealed in the depths of the human soul,* whose real modes of activity nature is hardly likely ever to allow us to discover, and to have open to our gaze.[26] (Italics added.)

Much of Kant's description could be accepted today, except that we would attribute this "concealed art" to the "depths" of the human brain rather than the soul.

Such schematic concepts, as almost all the concepts of the physical sciences, derive their sense from definitions in terms of more familiar or concrete words and meanings. Thus, contrary to both Berkeley's image theory of meaning and the positivists' verifiability criterion of meaning, whether a concept is meaningful does not depend upon being able to form a concrete image of it or upon the possibility of its being verified. As long as the definition consists of meaningful terms used in an intelligible grammatical structure, then the concept will be meaningful regardless of whether its referent exits or whether it can be imagined or empirically confirmed, such as Kant's notion of noumena or the concept of N-dimensional space in physics.

The role of definitions in constructing abstract concepts raises the question of the acquisition of language and its function in knowledge (again, these issues will be pursued further in the following two chapters). Language learning, especially, illustrates the existence of the following preconditions: (1) an innate ability to discriminate sensory qualities, forms, actions, and so on (possessed by all animals); (2) an inherited physiological system that causes a child to babble (not possessed by apes); (3) the reinforcing stimuli of the parent at the appropriate maturation period (not true of feral children); (4) the proper intercommunication of the dominant (with the speech center) and minor brain hemispheres (not true of split-brain patients); and (5) probably, a "species specific" innate "language-acquisition device" posited by Chomsky[27] (not possessed by animals). Given the natural physiological and maturational conditions, the child's instinctive babbling and imitative behavior, and the parents' typical reinforcing activities, any normal child learns to associate specific words or names with certain objects, actions, and qualities, and then to formulate these words in grammatically correct expressions.

Thus taking into account all the sufficient conditions, there is no mystery as to how words become "attached" to the world, nor is it necessary to appeal to such mystifying notions as "pictorial form," "logical form," "feelers," and "reaches right out to it," metaphors Wittgenstein used in the *Tractatus* to explain how language can express facts and represent states of affairs.[28] All animals have the physiological capacity to discriminate among sensory stimuli, while infants are genetically predisposed to learn that words stand for discriminated entities and therefore can be used in place of them in communication. As previously mentioned, chimpanzees have learned to use symbols and hand signs to communicate, the most talented, Washoe and Sarah, reputedly acquiring a vocabulary of 160 signs. Sarah, especially, possessed a remarkable language competence: "She used words, sentences, the interrogative, class concepts, negation, pluralization, conjunction, quantifiers, the conditional, and the copula."[29] Moreover, that the rate of their learning increased with training suggests that the cerebral cortex of chimpanzees may have an innate potential for learning to communicate by signs or symbols, so that the capacity to communicate might not be as "species specific" as Chomsky claims. Nevertheless, it is true that of the species on earth only human beings have been able to use symbols in so many marvelously innovative ways, from literature and physics to mathematics and music.

In addition to the empirical conditions already mentioned, communication has traditionally been analyzed into three components, as represented by the "language triangle": (1) physical signs (such as road signs, or dark clouds that are recognized as natural signs of rain), symbols (such as the national flag or the Christian cross), or words (spoken, printed, or written); (2) the meanings that signs, symbols, or words must have to be intelligible to a conscious organism; and (3) that which is being designated, referred to, or intended by the particular use of the signs, symbols, or words. (There are other aspects of the meaning situation, such as the emotive or subjective connotation of words, but these are less important for the present discussion, although crucial for understanding the literary or poetic function of language.) Each of the three elements of communication is clearly necessary. There must be a physical conveyer of the communication, such as spoken or written words; these linguistic symbols must have some meaning or nothing would be communicated (as was the case with Egyptian hieroglyphics before the discovery of the Rosetta stone in 1799); and, since the purpose of communication is to convey information about something, the signs and linguistic symbols must be understood to refer to or designate something. Originally, linguistic symbols, as well as most signs, derive their meanings from their referents by virtue of ostensive definitions (by pointing to the referent as the word is used), thereby acquiring, in Quine's terms, "stimulus meaning". But as languages develop and as an individual's language competence improves, stipulative definitions and contextually derived meanings based on the understanding of other terms within a written or verbal context become increasingly more important in learning the meanings of new words. Though generally ignored in philosophical discussions of language, reading is a primary means of increasing our vocabularies.

It has been customary among philosophers to designate the cognitive relation between the physical sign or word and its meaning, the "connotative" or "intensional meaning" (referring not to the individual subjective or emotive meaning but to the literal meaning consisting of a cluster of identifying attributes), while the referential function of words has been called the "denotative" or "extensional meaning." Thus the connotative or intensional meaning of 'whale' is the dictionary definition of 'whale' as a "large acquatic mammal of the order Cetacea," while the denotative or extensional meaning is the class of actual whales (whether the membership includes only living whales or also stuffed and literary whales is a matter of decision). Given these "modes of meaning," as they have been called, the problem of how language "attaches" to the world, or less metaphorically, how language can be used to refer to actual or possible states of affairs, as well as abstract entities, is not so difficult to understand after all.

In fact, the very conditions for learning a language satisfy the requirements necessary for its being used to designate phenomena. When a child learns to use such terms as 'dog,' 'brown,' and 'bark,' he or she has acquired some "criterion in mind" (C.I. Lewis's term), based on previous sensory discriminations, by means of which it is possible to identify and designate the animal dog, the color brown, and the action of barking. Similarly, an histology student not only has to learn the meaning or cluster of attributes belonging to terms such as 'pyramidal cell,' but also, to identify such cells under a microscope. Individuals could not be said to know the full sense or meaning of a word if they could define it but not correctly identify its referent, or identify its referent but not correctly define it (though occasionally we do not know both these modes of meaning). Thus the learning of a language, with its three essential components, word, meaning, and referent, *ensures that language can be used to refer to the world.*

As Wittgenstein maintained in the *Tractatus,* we use language to "project" actual or possible facts and/or states of affairs. In understanding a sentence (which Wittgenstein called a "propositional sign," the *physical* conveyer of the meaning or proposition itself), we consciously picture the fact or state of affairs referred to or projected by the statement (actually, we need not picture the state of affairs as such; instead, hearing or reading a statement such as "it is snowing outside" puts us in a state of expectation so that we are prepared to identify the referent as occurring or not). To quote from the *Tractatus*:

'A state of affairs is thinkable': what this means is that we can picture it to ourselves. (3.001)
The totality of true thoughts is a picture of the world. . . .(3.01)

In a proposition a thought finds an expression that can be perceived by the senses. (3.1)

We use the perceptible sign of a proposition (spoken or written, etc.) as a projection of a possible situation.
The method of projection is to think of the sense of the proposition [or more correctly, the propositional sign]. (3.11, brackets added.)

Unfortunately, however, Wittgenstein carried the notion of picturing too far. Instead of just maintaining that *we* picture states of affairs when we understand a printed or spoken sentence, he also claimed that the propositional sign "depicts" reality, analogous to hieroglyphic script.

At first sight a proposition [here again the context indicates a confusion between the proposition, or sense of a statement, and the physical conveyer of the proposition itself, namely the propositional sign] — one set out on the printed page, for example — does not seem to be a picture of the reality with which it is concerned. . . .(4.011; brackets added)

In order to understand the essential nature of a proposition [or better, propositional sign], we should consider hieroglyphic script, which depicts the facts that it describes.
And alphabetic script developed out of it without losing what was essential to depiction. (4.016; brackets added)

Perhaps, at one time, when language consisted of hieroglyphic script it could have been maintained that it literally pictured phenomena, but now that language has become so conventionalized, such an interpretation has little plausibility. Can it really be maintained that the three following sentences, with its different positioning of verb and direct object, literally picture reality? "I hit him"; "je l'ai frappé"; "Ich habe ihn geschlagen". If this were true, then reading any foreign language would be no more difficult than reading hieroglyphics. Yet Wittgenstein did maintain this view (though he subsequently gave it up), asserting that a propositional sign "like a *tableau vivant*—presents a state of affairs" (4.0311).
In addition, he asserted not only that *propositional signs* literally depict states of affairs, he also maintained that *propositions* themselves "picture reality." Just as non-abstract paintings represent scenes by virtue of their "pictorial form" so, Wittgenstein claimed, a proposition pictures reality by virtue of its "logical form."

What a picture must have in common with reality, in order to be able to depict it—correctly or incorrectly—in the way it does, is its pictorial form. (2.17)

What any picture, of whatever form, must have in common with reality, in order to be able to depict it—correctly or incorrectly—in any way at all, is logical form, i.e., the form of reality. . . .(2.18)

A proposition communicates a situation to us, and so it must be *essentially* connected with the situation.
And the connection is precisely that it is its logical picture. . . .(4.03)
Propositions can represent the whole of reality, but they cannot represent what they must have in common with reality in order to be able to represent it—logical form. (4.12)

But the notion of language itself picturing reality via a mysterious logical form, a form which could not be described by, but only "shown" in, language, explained very little, as Wittgenstein came to realize himself. It is *empirical investigations* in neurophysiology, psychology, and linguistics that explain how human beings learn to use language to refer to, describe, or picture the world, not a postulated "logical form". And though there is some similarity between Wittgenstein's notion of a logical form inherent in language and Chomsky's notion of a deep syntactic structure underlying all languages, the explanatory role of the two differ entirely.

To return to our discussion, while originally the development and learning of language depend upon the prior ability to discriminate among phenomena, once a language is acquired it has an overriding influence on how we categorize and describe the world. As the American linguist Benjamin Whorf has graphically stated:

We dissect nature along lines laid down by our native languages. The categories and types that we isolate from the world of phenomena we do not find there because they stare every observer in the face; on the contrary, the world is presented in a kaleidoscopic flux of impressions [an exaggeration, as we have seen] which has to be organized by our minds—and this means largely by the linguistic systems in our minds. We cut nature up, organize it into concepts, and ascribe significances as we do, largely because we are parties to an agreement to organize it in this way—an agreement that holds throughout our speech community and is codified in the patterns of our language.[30] (Brackets added.)

Thus one's conception of the world is conditioned by his cultural and intellectual milieu as transmitted by and reflected in language. While, as Chomsky claims, there could well be an inherited neurological mechanism accounting for a deep syntactic structure common to all languages, derived from certain evolutionary invariants such as the human physiological system, a commonly shared social organization, and a similar natural environment, there are great variations in the grammatical structures of natural languages, as described by Whorf. Thus the apparent contrast between the Whorf–Sapir theory of "linguistic relativity" and Chomsky's thesis of "a universal deep structure" underlying all languages will probably be resolved (as with the nature versus nurture controversy) in a more complex synthesis of both views—if this is not already implied in Chomsky's notions of deep and surface structures, and of transformational grammars.[31]

For the present study, what is important is the realization that personal knowledge (as compared to that recorded in books), which Popper calls World 2,[32] is stored in the human brain in the form of concepts or meanings which are constantly being elicited and manifested as we react intelligently to various environmental stimuli, while "the limits of language," as Wittgenstein asserted in the *Tractatus,* basically delineates, "the limits of. . .[one's] world" (see 5.62). The latter is not quite true because there seems to be another level of subcortical, subconscious experience that is the source of intuitive or metaphysical insight and aesthetically or mystically integrative experiences—the kinds of experience (such as those of Descartes and Pascal) that result in profound intellectual conversions. But granting the significance of these experiences attuned, perhaps, to some deep, ineffable domain, most of our knowledge is conceptual and expressed in the medium of one's language, since it is mainly through language (with all its possible symbolic variations) that we communicate information about ourselves, our acquired knowledge, and about the world. Thus language apart from experience would be sterile, while experience apart from language would be mute.

Notes to Chapter II

1. Cf. Robert L. Fantz, "The Origin of Form Perception," in R. Held and W. Richards (eds.), *Perception: Mechanisms and Models* (San Francisco: W. H. Freedman & Co., 1950), pp. 335–336.
2. Cf. Eckhard H. Hess, "Space Perception in The Chick," in R. Held and W. Richards (eds.), *op. cit.*, pp. 367–371.
3. Cf. Eleanor J. Gibson and Richard Walk, "The Visual Cliff," *ibid.*, pp. 341–348.
4. Cf. Irvin Rock and Charles S. Harris, "Vision and Touch," *ibid.*, p. 270.
5. Robert L. Fantz, *op. cit.*, p. 340.
6. Cf. T. G. R. Bower, "The Visual World of Infants," *ibid.*, pp. 355–357.
7. Cf. Steven Rose, *The Conscious Brain* (New York: Alfred A. Knopf, 1975), p. 153.
8. Cf. John C. Eccles, *Facing Reality* (New York: Springer–Verlag, 1970), pp. 66–68.
9. Cf. Jean Piaget, *The Origins of Intelligence in Children,* trans. by Margaret Cook (New York: International University Press, 1952), p. 3.
10. D. W. Hamlyn, *Experience and the growth of understanding* (London: Routledge & Kegan Paul, 1978), p. 100.
11. Ludwig Wittgenstein, *Philosophical Investigations,* trans. by G. E. M. Anscombe (Oxford: Basil Blackwell, 1953), Part I, sec. 33.
12. Cf. Joyce D. Fleming, "Field Report: The State of the Apes," in *Psychology Today,* ed. by T. G. Harris (New York: Ziff–Davis Publication Co.), January, 1974. Also, Eugene Linden, *Apes, Men, and Language* (New York: Saturday Review Press/E. P. Dutton & Co., 1974), chs. 2, 13.
13. Cf. Jean Piaget, *The Construction of Reality in the Child,* trans. by Margaret Cook (New York: Basic Books, 1954), ch. 1. Also, D. W. Hamlyn, *op. cit.,* p. 106.
14. For an excellent discussion of a "dispositional" conception of concepts and the various ways they are manifested, see H. H. Price, *Thinking and Experience* (London: Hutchinson's University Library, 1953), ch. XI.
15. Bernard Bosanquet, *Three Lectures on Aesthetics* (New York: Macmillan Co., 1915), p. 64.
16. Howard Gardiner, *The Shattered Brain* (New York: Vintage Books, 1976), p. 69.
17. Ludwig Wittgenstein, *op. cit.,* Part I, sec. 329.
18. Gilbert Ryle, *The Concept of Mind* (London: Hutchinson's University Library, 1949), p. 40.

19. W. V. Quine, *Ontological Relativity and Other Essays* (New York: Columbia University Press, 1969), p. 27.

20. Henry Mehlberg, *Time, Causality, And The Quantum Theory,* Vol. I (Dordrecht: Holland/Boston: D. Reidel Publishing Co., 1980), p. 284–285. Mehlberg supports this claim with an incident of problem solving by von Neumann.

21. Cf. W. V. Quine, "Mind and Verbal Disposition," in *Mind and Language,* ed. by Samuel Guttenplan (Oxford: Oxford University Press, 1975), p. 86.

22. W. V. Quine, *Ontological Relativity and Other Essays, op. cit.,* p. 26. Italics added.

23. Cf. Gerald Holton, *Thematic Origins of Scientific Thought* (Cambridge: Harvard University Press, 1973), pp. 367–372. Also, Einstein's "Autobiographical Notes," in *Albert Einstein: Philosopher-Scientist,* ed. by Paul A. Schilpp (Evanston: Library of Living Philosophers, 1949), p. 7ff.

24. Ludwig Wittgenstein, *op. cit.,* Part I, sec. 154.

25. John Hospers, *Understanding the Arts* (Englewood Cliffs, N.J.: Prentice-Hall, 1982), p. 36.

26. Immanuel Kant, *Critique of Pure Reason,* trans. by Norman Kemp Smith (New York: Humanities Press, 1929), pp. 182–183.

27. Noam Chomsky, *Language and Mind,* enlarged ed. (New York: Harcourt Brace Jovanovich, 1972), p. 113.

28. Cf. Ludwig Wittgenstein, *Tractatus Logical-Philosophicus,* trans. by D. F. Pears and B. F. McGuinness (London: Routledge & Kegan Paul, 1961), sec. 2.1511–2.22.

29. Joyce D. Fleming, *op. cit.,* p. 35.

30. Benjamin L. Whorf, *Language, Thought, and Reality,* ed. by John B. Carroll (New York: MIT Press and John Wiley & Sons, 1956), p. 213.

31. Cf. Noam Chomsky, *op. cit.,* p. 103ff.

32. Karl R. Popper and John C. Eccles, *The Self and Its Brain* (New York: Springer International, 1977), ch. P2.

CHAPTER III

CRITIQUE OF RECENT TRENDS IN PHILOSOPHY

IN 1963 Wilfrid Sellars wrote that the "trend in recent epistemology away from. . .classical phenomenalism ('physical objects are patterns of actual and possible sense contents') has become almost a stampede."[1] Since this was written, not only have epistemologists abandoned phenomenalism, they have abjured all of the epistemological principles of the positivists, along with most of the presuppositions of the ordinary language philosophers of mid–century. Today, the trend is away from a radical nominalism and instrumentalism in the philosophy of science toward a qualified essentialism and realism. Because this transition involves crucial questions pertaining to the explanatory function and truth value of scientific theories, which in turn involves issues pertaining to the reference and meaning of the words and sentences comprising these theories, it will be useful to review in perspective these earlier positions and the reasons for the changes, although the implications will be discussed in more detail throughout the book.

Logical positivism or empiricism was the dominant influence in American philosophy in mid–century, and its legacy is still so familiar that only a brief summary of its "presuppositional framework"[2] is necessary: the rational reconstruction of science on the

"rock bottom" (Carnap's term) foundation of indubitable sense data
with the help of logico–mathematical auxiliaries; promulgation of
the verifiability criterion of meaning, hence metaphysical statements were nonsense and ethical judgments were emotive; rejection of scientific realism, the view that the truth of most scientific
theories depends upon their approximate representation of unobservable domains of physical reality; acceptance of the doctrine that
all mathematical statements are analytic and the logistic position
that mathematical concepts are reducible to logic; and, finally, defense of the strict distinction between analytic and synthetic judgments and *a priori* and *a posteriori* knowledge, so that all
necessary knowledge must be analytic and *a priori* while all empirical knowledge must be synthetic and *a posteriori*. Due to the
extensive critical analysis of these tenets by the positivists themselves, as well as by their critics, none is accepted today.

Closely aligned with the positivists, at least as regards their
phenomenalism and empiricistic criterion of meaning, were pragmatists such as Dewey and C. I. Lewis, Dewey having contributed to
the "unity of science" publication of the positivists.[3] He also shared
the anti–metaphysical, anti–psychologistic, anti–dualistic orientation of the positivists, while his instrumentalist interpretation of
knowledge in general, and scientific theories in particular, was
congenial to the positivists, both having been influenced by the
revolutionary developments of relativity theory and quantum mechanics.[4] As regards relativity, Einstein acknowledged the influence
of Mach's positivism on his development of the special theory of
relativity, and endorsed operationalism when he subjected Newton's concepts of absolute space, time, and simultaneity to operational definitions, though later, he disavowed a positivistic
interpretation of scientific theories.[5] But it was the development of
quantum mechanics, especially in the early decades when it was
acknowledged that the mathematical formalism of Schrödinger's
wave equation and Heisenberg's matrix mechanics gave precise
statistical predictions of quantum processes without providing a
determinate description of the processes themselves, that supported an instrumentalistic interpretation of scientific theories.
This was especially true of the Copenhagen Interpretation of quantum mechanics, although Bohr was less of an instrumentalist and
more of a realist than Heisenberg.

Throughout this period, Wittgenstein, though often operating
behind the scenes, as it were, was extremely influential. His *Trac-*

tatus (1918), in which he presented the view that language, owing to its "logico-pictorial form," is a picture of reality while "thought itself is a logical picture of facts," had a marked effect on certain members, particularly Moritz Schlick, of the Vienna Circle. If natural science comprised "the totality of true propositions," as he maintained, philosophy could not compete as a separate body of doctrine and consequently was redefined as an activity of "elucidating propositions" and "clarifying thoughts" (4.11–4.12). Later, Wittgenstein disavowed the picture theory of language, though not the conception of philosophy as an analytical activity of clarification that was to have such a revolutionary effect on Anglo-American philosophy. In the privately circulated *Blue* and *Brown Books,* and especially in the *Philosophical Investigations* he, along with Gilbert Ryle, John Austin, and Peter Strawson, developed the technique of philosophizing that is often called "the Oxford School of ordinary language philosophy" (G. E. Moore was also a prominent ordinary language philosopher, though his mode of philosophizing was quite different from the Oxford School).

According to this School, philosophical problems have their origin in the misuse of ordinary language and therefore can be dissolved by showing that the language in which they are expressed consists of "systematically misleading expressions," "category mistakes," "logical howlers," or "linguistic muddles," in the words of Ryle, or can be attributed to the fact that "we do not command a clear view of the use of our words,"[6] as Wittgenstein put it.

The early sections of the *Philosophical Investigations* display Wittgenstein's struggle to rid himself of his earlier picture theory of language, offering a nominalistic, pluralistic conception as an alternative. "Family resemblances" replace essential definitions, and language is now thought of as consisting of a variety of "language-games," with the image of the toolbox substituted for that of the picture.

> Now what do the words of. . .language *signify?* — What is supposed to shew what they signify, if not the kind of use they have?. . . .Think of the tools in a tool-box: there is a hammer, pliers, a saw, a screw-driver. . .and screws. — The functions of words are as diverse as the functions of these objects (I, 10–11).

Usage not only determines signification, it constitutes meaning: "For a *large* class of cases. . .in which we employ the word 'meaning' it can be defined thus: the meaning of a word is its use in the language" (I, 43).

Wittgenstein thereby rejected the previous views of Mill and Frege that words have two modes of meaning, their extensional or denotative meaning derived from their referents, and their intentional or connotative meaning consisting of their associated cognitive criteria. He replaced these modes of meaning with the currently prevalent view that the meaning of words lies simply in how they are used—hence the origin and rationale of ordinary language analysis.

Thus the traditional conception of philosophy as an investigation of empirical problems that arise naturally in experience (for example, Descartes' formulation of the mind–body problem on the basis of such experiences as dreams and the phantom–limb illusion) or indirectly, as a result of empirical research and conceptual changes (as in Locke's distinction between primary and secondary qualities following the lead of Galileo, Newton, and Boyle), was transformed into linguistic analysis:

> ... Philosophical problems...are...not empirical problems; they are solved, rather, by looking into the workings of our language, and that in such a way as to make us recognize those workings: *in despite of* an urge to misunderstand them. The problems are solved, not by giving new information, but by arranging what we have always known. Philosophy is a battle against the bewitchment of our intelligence by means of language (I, 109).

Equally revolutionary was Wittgenstein's attack on inner experiences, mental processes, and private languages, as indicated in the previous chapter. Ever since Descartes, philosophers have assumed that what can be known with greatest certainty are our sense impressions and ideas, so that any reference to an independent external world must be inferred from these indubitable mental contents. Until Wittgenstein (though he was not alone in undermining this presupposition), philosophers as diverse as Russell, Moore, Broad, Lovejoy, Carnap, Lewis, Price, and Ayer all believed that knowledge is grounded in these immediate sensory data, "knowledge by acquaintance," as Russell called it.[7] Moreover, they assumed that each of us has a "privileged access," in Ryle's term, to such inner experiences as pains, feelings, meanings, and mental processes in general: that we "directly perceive" sense data, not physical objects, that only the person himself can know whether he or she is in pain, and that understanding is a mental process.

As deeply entrenched among philosophers as these notions had become, Wittgenstein found them objectionable because they conflicted with our common sense views and ordinary ways of speaking (hence the frequent reference among ordinary language philosophers to how "the plain man" thinks and speaks). They seemed to be puzzles peculiar to philosophers and hence were suspect. Wittgenstein maintained that philosophers get ensnared in such problems by inadvertently misusing ordinary language. Consider, for example, what he says regarding the position he had held earlier in the *Tractatus,* that the meaning of a word is the object it names.

> . . . This is connected with the conception of naming as, so to speak, an occult process. Naming appears as a *queer* connection of a word with an object.—And you really get such a queer connection when the philosopher tries to bring out *the* relation between name and thing. . .For philosophical problems arise when language *goes on holiday.* And *here* we may indeed fancy naming to be some remarkable act of mind, as it were a baptism of an object. And we can also say the word "this" to the object, as it were *address* the object as "this"— a queer use of this word, which doubtless occurs only in doing philosophy (I, 38).

Thus it is when doing philosophy that one is misled into thinking that some special "act of mind" is involved in learning the name of an object. So much for the problems of semantics!

Or consider Wittgentein's treatment of the question whether, when we are in pain, we can know this in a way that no one else can know it.

> In what sense are my sensations *private?*—Well, only I can know whether I am really in pain; another person can only surmise it.—In one way this is false, and in another nonsense. If we are using the word "to know" as it is normally used (and how else are we to use it?), then other people very often know when I am in pain.—Yes, but all the same not with the certainty with which I know it myself!—It can't be said of me at all (except perhaps as a joke) that I *know* I am in pain. What is it supposed to mean—except perhaps that I *am* in pain (I, 246)?

There are so many compelling counter-examples to this statement that one wonders how anyone could have asserted it in the first place, and why others took it seriously! Is there any person who has not been asked by a dentist or doctor, "does this hurt?" The very posing of the question presupposes a knowledge on the part of the patient that neither the dentist nor the doctor has access to. Is there a parent or a teacher who has not wondered on occasion whether a child is really sick or only pretending? If so, doesn't this imply a doubt on the part of the parent or teacher that is not true of the child? Can the witnesses to an electrocution know in the way that the condemned person does what the final death trauma is? Any claim to know something presupposes a description of what one knows, but can anyone else describe the pain an amputee or a cancer patient suffers, as well as that person? Enlarging the scope of "private" experiences, how can one deny that only the person having the horrifying nightmare, the terrifying illusion (delirium tremens or drug induced hallucinations), the severe depression, or the debilitating feeling of persecution suffered by a schizophrenic can truly know what he or she is experiencing?

Nonetheless, Wittgenstein thought he had disposed of the privateness of inner experiences by showing that the peculiar sense of uniqueness was not a function of the experience but of the language used to express or describe it.

> "I know. . . .only from my *own* case"—what kind of proposition is this meant to be at all? An experiential one? No.—A grammatical one?
> Suppose everyone does say about himself that he knows what pain is only from his own pain.—Not that people really say that. . .But *if* everyone said it. . .And even if it gives no information, still it is a picture, and why should we not want to call up such a picture?. . . .
> When we look into ourselves as we do philosophy, we often get to see just such a picture. A full–blown pictorial representation of our grammar. Not facts; but as it were illustrated turns of speech (I, 295).

As long as we use language in the usual way we convey facts, but as soon as we lapse into a philosophical use of language by asking certain kinds of questions, we conjure pictures which are representations of grammar rather than facts.

This notion is clearly illustrated in Wittgenstein's analysis of the 'theory' of sense data. The 'theory' is not considered a hypothesis to account for the difference between the qualitative world as we experience it because of our particular sense organs and the physical world as conceived by the physicist, but is depicted as another false way of picturing things owing to the philosopher's peculiar use of language. (Wittgenstein's influence has been so pervasive that many philosophers have lost all confidence in the possibility of formulating philosophical theories. When discussing the cluster theory of names, for example, Saul Kripke says: "It really is a nice theory. The only defect I think it has is probably common to all philosophical theories. It is wrong.")[8]

> You have a new conception and interpret it as seeing a new object. You interpret a grammatical movement made by yourself as a quasi-physical phenomenon which you are observing. (Think for example of the question: "Are sense-data the material of which the universe is made?" (I, 401).

As a final illustration of Wittgenstein's method of dealing with traditional philosophical problems (also discussed in the previous chapter), consider the mind-body problem. Neurophysiologists and neuropsychologists have amassed considerable evidence that our conscious processes depend upon extraordinarily complex brain structures and interactions consisting of nerve firings transmitting chemical-electrical discharges. But how do these intricate neuronal processes give rise to or come to be experienced as ordinary conscious processes of feeling, perceiving, imagining, remembering, and thinking? The tremendous difference between the descriptive predicates and modes of access to the two has aroused the bewilderment of both scientists and philosophers, yet Wittgenstein sees the problem as another instance of misusing language.

> The feeling of an unbridgeable gulf between consciousness and brain-process: how does it come about that this does not come into the considerations of our ordinary life? This idea of a difference in kind is accompanied by slight giddiness,—which occurs when we are performing a piece of logical slight-of-hand. ...When does this feeling occur in the present case? It is when I, for example, turn my attention in a particular way on to my own consciousness, and, astonished, say to myself: THIS is supposed to be produced by a process in the brain!—as it were clutching my forehead. ...
> Now bear in mind that the proposition which I uttered as a paradox (THIS is produced by a brain-process!) has nothing paradoxical about it. I could have said it in the course of an experiment whose purpose was to shew that an effect of light which I see is produced by stimulation of a particular part of the brain.—But I did not utter the sentence in the surroundings in which it would have had an everyday and unparadoxical sense (I, 412).

This is a strange argument, implying that because in our everyday life we are not perplexed as to how conscious processes are related to brain processes, we should not be perplexed at all! But most of our theoretical questions, whether philosophical or scientific, would not arise if we merely attended to the usual occurrences of phenomena, ignoring their causes or conditions. What is perplexing about the fact that unsupported physical objects fall or that a bolt of lightning strikes during a thunder storm? As long as our brains function normally, we can speak and comprehend others in an effortless way—but when these processes are impeded by a stroke, we begin to wonder about their underlying causes. As long as one attends only to macroscopic objects, no question arises as to how they can consist of molecules, atoms, and nuclear particles and forces, as we now believe. When we look at the night sky it appears as if our vision reaches across space to make contact with the stellar bodies, and yet we know that what we see is due to the immediate stimulation of light waves or photons on our sense organs. In his attempts to dispel philosophical problems, Wittgenstein, along with other ordinary language philosophers, conveyed the impression that the "plain man" is somehow more intelligent than the philosopher because he goes through life relatively unperplexed by certain problems. Recall G. E. Moore's statement that "the sciences would not have suggested to me any philosophical problems."

Wittgenstein's method of dealing with typical problems of philosophy has been illustrated in some detail because of his enormous influence on Anglo–American philosophers. He, more than anyone else, is responsible for the "linguistic turn" in philosophy. It is largely owing to him that the meanings of words are now usually equated with their use, while his emphasis on "external criteria" when assessing one's intellectual performance redirected philosophy along behavioristic lines, as illustrated in Ryle's *The Concept of Mind* which followed out the implications of many of Wittgenstein's assertions in the *Philosophical Investigations*—just as Austin took his cue from Wittgenstein's distinction between the use of 'know,' 'can,' and 'is able to' (I, 150).

Merely describing Wittgenstein's method of investigation does not, of course, constitute a critical evaluation or a refutation of his views, although I have indicated some points of disagreement along the way. My purpose has been to show that his approach is, after all, based on certain predispositions and assumptions as to the nature of philosophical perplexity, and is not the final solution to these problems, as his early followers tended to believe. As I shall attempt to illustrate throughout the book, the traditional problems of philosophy are too basic and empirically grounded to be "dissolved" by linguistic analysis. Now that the implications of Wittgenstein's conception of philosophy and how it should be pursued are better understood, it is hoped that philosophers will turn to more effective alternatives. If, after all, philosophical problems *were* nothing more than spurious grammatical pictures, then the most direct way of eliminating them would be by discontinuing the teaching of philosophy—why contaminate more students?—a consequence that ordinary language philosphers were not up to facing, although their conception of philosophy is largely responsible for its current eclipse.[9]

While the positivists had dethroned metaphysics as the queen of the sciences, accepting science itself as the only rightful pretender to the realm of knowledge, and Wittgenstein had disposed of the claims to legitimacy of epistemological problems by attributing them to grammatical confusions, other forces were at work undermining the rational sovereignty of science itself. In opposition to the positivists' attempt to ground the edifice of science on the "rock bottom" foundation of sensory evidence shored up by logico–mathematical structures, Popper denounced the principles of induction and verification, arguing that the only valid methodological principle was that of falsification. Accepting Hume's critique of induction, that repeatable instances are not valid grounds for future expectations, along with the fact that no amount of confirmation of the predictable consequences of a theory would prove it true, Popper concluded that the only correct procedure was a rigorous effort to falsify theories. Theories cannot be known to be true, only to have not yet been falsified, thus "verisimilitude' or approximation of the truth is all we can expect: "the aim of science is truth in the sense of better approximation to truth or greater verisimilitude."[10] But Popper's methodological principle not only inverted the actual procedure of scientists (since normally they attempt to confirm theories, not disprove them), his efforts to define verisimilitude in terms of "high truth content" and "low falsity content" turned out to be less than convincing.

Moreover, the conception of falsification itself contained an inherent problem regarding the status of the falsifying data. If, as Norwood Hanson argued (again showing the influence of Wittgenstein), all knowledge is "theory laden"[11] in the sense that no observations are independent of some conceptual–linguistic interpretation, then observations cannot be decisively falsifying. When confronting disconfirming instances, one always has the option of reinterpreting the data to save the theory, thus theories are no more conclusively disconfirmed than confirmed, according to this view (though I think this is somewhat exaggerated if one considers actual instances, such as the Michelson–Morley experiment).

The final challenge to science's claim to the sovereignty of rationality was posed by Thomas Kuhn. In his iconoclastic analysis of science, *The Structure of Scientific Revolutions,* Kuhn denied the traditional view of science as a gradual but steady progression toward a true theory of the world, insisting that a closer examination of the history of science discloses not only relatively smooth progressions (normal science), but discontinuous transitions (revolutionary science). In these revolutionary changes, not only are contesting theories "incompatible" or "incommensurate," they involve cognitive processes and modifications that are neither rational nor intelligible: that is, sudden, irreversible gestalt switches from one paradigm to another (based partially on irrational factors) that transform the scientist's cognitive structure so radically that he literally finds himself in a different world.[12]

Also basing his conclusions on detailed analyses of scientific changes, Paul Feyerabend went beyond Kuhn in embracing an "epistemological anarchism." Arguing that adherence to *any* fixed principles of scientific method would have impeded scientific progress because it can be shown historically that certain revolutionary developments occurred precisely because their proponents ignored the accepted rational evidence and arguments, often resorting to "a mixture of subterfuge, rhetoric and propaganda," Feyerabend adopts the principle that "anything goes."[13] Moreover, he maintains that science has no valid claim of superiority over any other discipline, be it voodooism, witchcraft, or astrology. Ostensively, the heuristic value of this position is that it would promote a tolerance for all contending views, and because Feyerabend believes that uncovering significant new empirical evidence depends on adopting radical alternatives to accepted theories, it would promote the advance of knowledge.[14]

There have, of course, been rejoinders to these extreme positions that will be addressed either directly or by implication later. However, a brief indication of the reservations one has to these views is appropriate now. For example, there has been considerable criticism of Kuhn's conviction that paradigm changes involve arational or irrational gestalt switches. Historical evidence does not indicate that there was such a sudden and irreversible change in Galileo's thinking, for example, but rather a gradual transition in outlook from the time he taught Aristotelian science at the University of Padua. Similarly, there is no evidence that such a change occurred in Copernicus or Einstein. Even Planck's acceptance of a discontinuous, irreducible quantum of interaction, in contrast to continuous physical processes, was not due to a sudden gestalt switch, as far as we know, but rather to an extremely reluctant acceptance of the fact that the Rayleigh–Jeans' or Wein's law was incorrect, and that "the ultraviolet catastrophy" could be accounted for only by positing *discrete* energy exchanges. Moreover, the fundamental *justification* of these conceptual shifts was based on *rational arguments.* If theories were in fact as incompatible or incommensurable as Kuhn claimed, there would be no basis at all for comparing them or for claiming that one was, at least in certain respects, superior to the others.

As for Feyerabend, while his principle of tolerance is admirable, it is difficult to understand how science could have progressed if there had not been some constraint on the kinds of evidence and arguments that were deemed acceptable. Even if, as certainly was the case, those scientists most responsible for revolutionary changes often adhered to and advanced their views despite contrary evidence and accepted rational arguments, this does not mean that their own thinking was irrational, but rather, that it took account of evidence and rationale *not yet* convincing to others. Scientific revolutions do bring about radical modifications in our basic assumptions and patterns of reasoning, but the acceptance of these changes was not determined by a mixture of "subterfuge, rhetoric and propaganda," but by a rational assessment of the evidence. The case of Lysenko and the disclosure of fraudulent scientific claims refute Feyerabend's argument that scientific positions can be maintained on irrational grounds. In addition, although all scientific theories are fallible in the sense that they are subject to falsification or revision, only an artful stubbornness would prevent one from acknowledging that our present understanding of genetics has been tremendously advanced as a result of the discovery of the double helix structure of DNA, and that our current conception of the solar system is far superior to that of Anaximander.

The last philosopher to be considered is an American, and though he has not had the revolutionary impact of Wittgenstein, his influence has been so pervasive that today his views are more widely discussed. Unlike Wittgenstein who, Quine says, "found a residual philosphical vocation. . .in curing philosophers of the delusion that there were epistemological problems,"[15] he is concerned with rehabilitating epistemology following the attacks on it by ordinary language philosophers and the collapse of the research program of the logical positivists. In fact, having contributed decisively to the rejection of two major tenets of the positivists—the "fundamental cleavage" between analytic and synthetic truths and the notion "that each meaningful statement is equivalent to some logical construction upon terms which refer to immediate experience"[16]—Quine would like to salvage what was true in the positivists' program in his effort to "naturalize" epistemology.

Accepting the impossibility of grounding science on some prior, absolute epistemic foundation, Quine views epistemology as a semantic or linguistic inquiry, not in Wittgenstein's sense, but with the goal of showing how, on the initial basis of the stimulation of our sense organs, we are able to arrive at our theory of the world, which for him means *talk* about the world. As Roger Gibson says in his excellent book on Quine: "Quine reformulates the question of how we acquire our theory of the world as the question of how we acquire our *theoretical talk* about the world. . .because he regards theories, from first to last, to be linguistic structures."[17] That Quine's inquiry is a semantic one, concentrating on how people, scientists included, come to *talk* about the world as they do is crucial because it affects his entire orientation. Because of this he finds epistemology especially affiliated with psychology and focuses on such problems as language learning, indeterminacy of translation, and inscrutability of reference, rather than on the role of experimentation in scientific inquiry, theory construction, or the significance of conceptual changes in science—the concerns of philosophers of science. As he says, "our objective. . .is. . .a better understanding of the relations between evidence and scientific theory. Moreover, the way to this objective requires consideration of linguistics and logic along with psychology."[18] This is not the way a philosopher of science would construe that objective!

In this way Quine can be seen as attempting to replace the reductivistic program of the positivists with a semantic inquiry, discarding their conception of scientific knowledge as a deductive, static structure of statements for a genetically explained, flexible linguistic network, *without,* however, *discarding the observational foundation.* As he says:

It was sad for epistemologists...to have to acquiesce in the impossibility of strictly deriving the science of the external world from sensory evidence. Two cardinal tenets of empiricism remained unassailable, however, and so remain to this day. One is that whatever evidence there *is* for science *is* sensory evidence. The other, to which I shall recur, is that all inculcation of meanings of words must rest ultimately on sensory evidence (OR, 75).

However, as we shall find, neither tenet is unproblematic. The evidence that science now relies on is derived mainly from experimentation rather than sensory observation, with theories often overdetermining the data. Even the "inculcation of meanings of words" is different from that envisioned by Quine, as I shall attempt to show.

Turning now to his total philosophical system, Quine's gift for colorful linguistic expression makes for easy classification of his essential tenets: indeterminacy of translation, inscrutability of reference, linguistic holism, underdetermination of theories by empirical evidence, and ontological relativity. Each of these theses, in turn, is dependent upon his naturalistic–behavioristic framework. As Gibson states,

Quine's commitment to behaviorism may very well prove to be his Achilles' heel; for if, at some time in the future behaviorism (in Quine's sense of the term) is abandoned, say, in favor of some form of mentalism, then an overwhelming majority of Quine's cardinal doctrines and theses would be left without any obvious firm support and, therefore, cast into doubt.[19]

Although I have criticized Quine's attack on "mentalism" in the previous chapter, since his philosophical system "hinges on his behavioristic orientation," as Gibson reiterates, it deserves further consideration.

In his important John Dewey lectures at Columbia University, entitled "Ontological Relativity," Quine describes the rationale for his naturalistic–behavioristic commitment. Claiming continuity with Dewey, and certainly agreement with Wittgenstein's behavioristic orientation in the *Philosophical Investigations,* Quine stresses his "naturalistic" denial of meanings as mental entities or private psychic states (though why this should be considered "naturalistic" rather than reductivistic or physicalistic, he does not say). This denial, in turn, is based on his belief that an explanation of language learning and use can be given in purely behavioristic terms; thus, he rejects what he derisively calls "the myth of a museum in which the exhibits are meanings and the words are labels" (OR, 27), a rejection that underlies his doctrine of the indeterminacy of translation. "When with Dewey we turn thus toward a naturalistic view of language and a behavioral view of meaning, what we give up is not just the museum figure of speech. We give up an assurance of determinacy" (OR, 28).

Although it is generally accepted outside of philosophy that the difference between understanding and not understanding words, and being capable or incapable of translating from one language to another, depends upon our knowing or not knowing the 'meanings' of the words, these meanings being the intelligible contents of conscious states (notwithstanding their dependence on neurological processes), Quine, like Wittgenstein, denies this. Nothing is gained, he thinks, by presuming to explain speech "by appealing to mind, mental activity, and mental entities: by appealing to thoughts, ideas, meanings"[20]—the "idea idea" as he sometimes calls it. His objection rests on the conviction "that language is a social enterprise which is keyed to intersubjectively observable objects in the external world" (p. 84) that, as children, we learn "on the evidence solely of other people's overt behavior under publicly recognizable circumstances" (OR, 26). Like Wittgenstein, he believes that the meanings of words consist in their various *uses,* in one's *overt use of speech.* Understanding a word does not consist of a light bulb clicking on in our mind, but of our being able to produce "an equivalent expression that is clearer" (p. 86).

For Quine, then, mental activity can be explained by our *use* of words, rather than by our possession of meanings, concepts, or ideas.

> People persist in talking thus of knowing the meaning, and of giving the meaning, and of sameness of meaning, where they could omit mention of meaning and merely talk of understanding an expression, or talk of the equivalence of expressions and the paraphrasing of expressions. They do so because the notion of meaning is felt somehow to explain the understanding and equivalence of expressions. We understand expressions by knowing or grasping their meanings; and one expression serves as a translation or paraphrase of another because they mean the same. It is of course spurious explanation, mentalistic explanation at its worst (pp. 86–87).

Why is reference to meanings in attempting to explain our understanding of an expression spurious explanation? The answer seems to lie simply in Quine's preference for behaviorism and in his belief that we think with words.

> The behavioural level. . .is what we must settle for [since we do not yet have neurophysiological explanations] in our descriptions of language, in our formulations of language rules and in our explications of semantical terms. It is here, if anywhere, that we must give our account of the understanding of an expression, and our account of the equivalence that holds between an expression and its translation or paraphrase. These things need to be explained, if at all, in behavioural terms: in terms of dispositions to overt gross behaviour (p. 87; brackets added).

One should note that Quine does not give any reasons why references to meanings are "spurious explanations," as common as they are, merely asserting that we "must settle for" behaviorism. Why must we settle for this? Could it be that behaviorism was so prevalent at the time due to Wittgenstein, Ryle (in the above quotation Quine uses the term "disposition," a favorite term of Ryle), and Skinner that Quine could assume no justification was necessary?

This is an example of one of the most curious and puzzling phenomena in philosophy, namely, that a position definitely preferable to some philosophers is, to others, obviously inadequate. How can an intellectual person, especially, find explanations of learning, talking, and theorizing limited to *external behavior,* sufficient, ignoring completely the internal cognitive efforts involved? One can accept everything Quine says about "language being a social enterprise" learned, at least partially, "on the evidence of other people's overt behaviour," but object that such a view leaves out the crucial *inner conscious processes* involved in learning. As scholars, much of our intellectual development has taken place in the quiet of our studies, isolated from other people, *thinking* about what we are reading or writing and *reflecting* on problems. Once adept at reading, we are no longer dependent upon the "overt behaviour of others," acquiring new meanings from definitions and explanations in scholarly texts and reference books. Yet Quine maintains that "even in the complex and obscure parts of language learning, the learner has no data to work with but the overt behavior of other speakers" (OR, 28). It is significant that he never once mentions books as a source of meanings, as if *talking* were the only activity relevant to learning. While books as a source of information obviously have an objective status, the assimilation and comprehension of the reading material is a conscious activity, not a behavioral one, unless one arbitrarily construes concentrating, comprehending, and reflecting as behavior.

These typical conscious endeavors are not equivalent to external conditioning and reinforcement but involve conceptualizing and thinking in which inchoate images, implicit concepts or incipient meanings, as well as words, play a necessary role. How often in private reflection, serious conversation, or when listening to a lecture do we find explicit words passing through our minds? How can "searching for the right word," "thinking out a problem," "intending to say something clearly," or "deliberately trying not to embarrass someone by what one says" be explicated in behavioral terms? In each case, the activity, along with the criterion of success, Wittgenstein notwithstanding, is mental, not behavioral, in the sense that the person is *aware* of whether he has succeeded or not, though behavioral cues may also help. While writing one's thoughts and calculating with a pencil normally accompany our efforts to clarify concepts and solve mathematical problems, the behavior of writing is not the thought process but its manifestation. Recall that in such endeavors we often suspend the physical action when we become engrossed in thinking about the problem. Or consider the role of *gedankenexperiments* in Einstein's development of relativity theory and in his discussions with Bohr (the EPR controversy) regarding the incompleteness or completeness of quantum mechanics, where mental images play a vital part. Recently, I took a course in computer operations and had the frustrating experience of being able to "drive the computer" according to the instructor's directions when she was present, but often being unable to replicate the behavior on my own because I did not understand the reasons for my initial operations. These are the inner conscious processes that drive much of human behavior. Even though neurological structures underlie our talking and thinking, on the conscious level these activities involve meanings and concepts, not neurological processes. Try thinking about a theoretical problem in philosophy in neurological terms and see how far you get!

In what has been written I realize that I have appealed to introspection and accepted the significance of such words as 'intending,' 'comprehending,' 'thinking,' and 'reflecting' which themselves are mentalistic terms,[21] but even Quine, in his most recent writing (in marked contrast to what he wrote in OR), states that such terms are *indispensable:* "Anyone, however tough his mind, is well advised to recognize both the irreducibility and the indispensability of personal history and the mentalistic idiom."[22] Much to his credit, in my opinion, is this latest evidence of Quine's continuous modification of his earlier more radical and uncompromising naturalistic–behavioristic position. But if the "mentalistic idiom" is "irreducible and indispensable," then one must confront the fact and incorporate the consequences in one's interpretation of how language is learned and used!

Quine's behavioristic thesis quoted earlier, that "language is a social art which we all acquire on the evidence solely of other people's overt behavior under publicly recognizable circumstances," seems not only too restrictive to accommodate the internal cognitive processes essential to most learning, it may also be refuted by recent investigations of creole languages. Quine and Noam Chomsky have long been engaged in a controversy whether Quine's somewhat Skinnerian behaviorist view is adequate to account for what Chomsky considers the three essential characteristics of human language competency mentioned in the last chapter: that the use of language be innovative in producing new expressions; that language be free from stimulus control in order that it can be used creatively for individual self–expression; and that it be "appropriate to the situation," in contrast to mere parroting or when there has been damage to the Wernicke's area of the brain (described earlier).

According to Chomsky, we cannot account for these uniquely human linguistic abilities merely on the basis of behavioristic interpretations utilizing reinforcement and conditioning. While verbal reinforcement is necessary for the child to learn a particular language, it is not sufficient to account for this learning which, Chomsky claims, is dependent upon a biologically endowed, "species–specific" faculty of language acquisition.[23] This innate species–specific faculty consists of deep grammatical structures (with restrictive principles) that enable any normally endowed child to transform the phonetic patterns and surface structure of its particular natural language into a meaningful linguistic system that in turn provides for the understanding and formulation of an endless variety of new linguistic expressions. Thus Chomsky finds Quine's account of language learning, like Skinner's, inadequate.

However, as a result of their continuing dialogue the gap between their positions has narrowed considerably. Chomsky has always acknowledged the necessity of verbal reinforcement in lan-

guage learning and admitted that his innate mental faculty may
ultimately be explained in neurophysiological terms,[24] while Quine
has acknowledged the necessity of positing, along with an innate
endowment for mimicry, an instinct of babbling, and innate stan-
dards of perceptual similarity, "some further innate apparatus which
is not yet identified."[25] However, while Quine, like Chomsky, concedes the need for some innate endowments to facilitate language
learning, he previously refused to think of them in mentalistic terms
or as cognitive processes. Such innate endowments are

> hypothetical mechanism[s], and our behavioral ticketing of them is
> only a partial description by a superficial, conditionally observable
> manifestation. Further mechanisms may be yet more remotely the-
> oretical, deep in the DNA; and I am just as receptive to such conjec-
> tures, regarding the mechanisms of language, as I am to the
> physicists' conjectures regarding elementary particles. I ask only that
> the positing of such "intervening variables" be motivated by an ear-
> nest quest for an eventual explanatory model, integrated with our
> overall physical system of the world. Such a prospect or project, and
> not mere affection for time–worn mentalistic myths, should be the
> motivation.[26] (Brackets added.)

I do not understand how this assertion can be reconciled with
his more recent statement regarding "the irreducibility and indispensability of. . .the mentalistic idiom."

Again, however, while we can agree that language learning is
dependent upon neurological mechanisms so that much of the
process takes place unconsciously, nothing we know about these
mechanisms as yet can help us understand or facilitate our *conscious* efforts to learn. No foreign language teacher, for example,
can help students learn a new language by directing their attention
to their neurons. This is why taking account of the cognitive elements involved, such as meanings, is essential. One does not just
learn how to use foreign words in the approriate context or on the
right occasion (for example, "*ca va?*"), but one learns their meanings, which is what one understands when the language is used. The
behavioral evidence that one understands consists of being able to
express the foreign statement in an equivalent expression in one's
own language (the ability stressed by Quine), but this does not
constitute the person's *understanding* of the expression; it presupposes it. Recall how often we have heard someone state, "I know
what I want to say, but I can't find the words."

The evidence in favor of an innate linguistic structure facilitating an infant's language learning has received strong support recently from the study of diverse creole languages (not to be confused with the language spoken by many blacks in southern Louisiana). These creole languages were developed throughout the world by the children of polyglot immigrant migrant workers whose diverse ethnic and linguistic backgrounds forced them to communicate in a rudimentary speech system called pidgin. The offspring of these immigrants, though mainly exposed to the impoverished pidgin speech of their parents and other members of the community without any common linguistic heritage or reinforcement, nevertheless developed a language which was not the pidgin of their parents, but a language *with a uniform structure that did not resemble that of any other language.* The vocabulary varies with the ethnic background of the particular group of creole speaking children, but the grammatical structure is strikingly uniform. As Derek Bickerton states:

> The implications of these findings are far-reaching. Because the grammatical structures of creole languages are more similar to one another than they are to the structures of any other language, it is reasonable to suppose most if not all creoles were invented by children of pidgin-speaking immigrants. Moreover, since creoles must have been invented in isolation, it is likely that some general ability, common to all people, is responsible for the linguistic similarities.[27]

In addition to the unexpected grammatical uniformity of diverse creole languages, Bickerton finds further evidence supporting the innateness hypothesis reflected in the grammatical mistakes children in any linguistic community are likely to make. These depend, Bickerton found, on the dissimilarity or similarity between the grammatical structure of creole and that of the local language. As one would expect, there are few errors when the (innate) structure of creole is similar to that of the local language whereas there is a marked resistance to certain speech patterns when the structures are different, the child preferring the (innate) creole structure.

While obviously contrary to Quine's behavioristic doctrine that language is learned "on the evidence solely of other people's behavior," this evidence demands a modification of Chomsky's nativistic thesis as well. Whereas Chomsky's postulated deep universal grammar consists of sets of diverse linguistic rules from which the child must select the one that conforms to his local language, the *uni-*

form grammar of creole languages indicates a more *specific* innate grammatical structure. As Bickerton asserts in the same article:

> The evidence from creole languages suggests that first–language acquisition is mediated by an innate device of a rather different kind [from Chomsky's]. Instead of making a range of grammatical models available, the device provides the child with a single and fairly specific grammatical model. It was only in pidgin–speaking communities, where there was no grammatical model that could compete with the child's innate grammar, that the innate grammatical model was not eventually supressed. The innate grammar was then clothed in whatever vocabulary was locally available and gave rise to the creole languages heard today (p. 121; brackets added).

Although he acknowledged the role of actual or possible innate endowments, the fact that Quine limited his explanation mainly to stimulation of sensory receptors and behavioral reinforcement predetermined his account of language learning, along with his notions of "indeterminacy of translation" and "inscrutability of reference." As a holist, Quine accepts sentences over words and even "blocks of theories" over sentences as the basic semantic units, yet the child must initially learn the meanings of terms by ostensive definition. Here especially, the verbal behavior of others *is* critical in reinforcing in the child the association of certain sounds or words with the appropriate referents when both the child and the reinforcer are confronted with the same sensory stimuli. The consequent matching of words with receptor stimulation Quine calls "stimulus meaning," the meaning characteristic of "observation sentences," those indispensable statements in his semantic theory that constitute the entry into language, the basis of any linguistic agreement, and the evidential support of science.

> The observation sentence is the cornerstone of semantics. For it is...fundamental to the learning of meaning. Also, it is where meaning is firmest. Sentences higher up in theories have no empirical consequences they can call their own; they confront the tribunal of sensory evidence only in more or less inclusive aggregates. The observation sentence, situated at the sensory periphery of the body scientific, is the minimal verifiable aggregate; it has an empirical content all its own and wears it on its sleeve (OR, 89).

Surprisingly, since even the reports of sensory stimuli could be stated or conveyed differently in different languages, Quine maintains that the "predicament of the indeterminacy of translation has little bearing on observation statements" (OR, 89), applying essentially to theoretical statements. But if observation sentences, unlike theoretical statements, are not indeterminate in translation because their meaning and evidence tend to coincide, then indeterminacy of translation, as well as inscrutability of reference, could be avoided, at least as regards these statements. To be consistent, Quine should have maintained that even observation sentences, as *sentences* (despite their closeness to receptor stimulation), can vary from one language to another, if the languages are sufficently diverse. But Quine's commitment to empiricism (his last link with positivism) prevented his taking this step: "The stimulation of his sensory receptors is all the evidence anybody has had to go on, ultimately, in arriving at his picture of the world" (OR, 75). Yet this statement is ambiguous because it is not bare stimulation of the senses that constitutes evidence, but this stimulation as interpreted within a cognitive–linguistic framework. As Paul Churchland asserts:

> The conviction that the world instantiates our ordinary observation predicates cannot be defended by a simple appeal to the "manifest deliverance of sense". *Whether or not the world instantiates them is in the first instance a question of whether the theory which embeds them is true.*...we may be systematically *mis*perceiving reality in the first place.[28]

In any case, once the child learns different strings of observation sentences, it can generate further variations of these by "analogical synthesis," a process of substituting similar terms for their analogues in the appropriate syntactic positions. Although the following example is his only illustration, it does indicate Quine's attempt to explain, without recourse to an innate language device, how the child acquires a repertoire of novel sentences.

> Most sentences are built up...from learned parts, by analogy with the way in which those parts have previously been seen to occur in other sentences....What sentences are got by such analogical synthesis, and what ones are got directly, is a question of each individual's own forgotten history.
>
> It is evident how new sentences may be built from old materials and volunteered on appropriate occasions simply by virtue of the analogies. Having been directly conditioned to the appropriate use of 'Foot' (or 'This is my foot') as a sentence, and 'Hand' likewise, and 'My foot hurts' as a whole, the child might conceivably utter 'My hand hurts' on the appropriate occasion, though unaided by previous experience with that actual sentence.[29]

As Quine himself admits, supposing this *were* an important factor in language learning, it still would account at most for an increasing variation of observation sentences, while what we want is a whole network of sentences constituting our natural languages and scientific theories. In *The Roots of Reference,*[30] he supplements this explanation with more detailed examples of how he thinks the various parts of language could be learned, but since the evidence of creole languages provides a more factual explanation, whereas Quine's is merely conjectural, there is no need to pursue further Quine's speculative account. Suffice it to say that eventually there emerges "a fabric of sentences variously associated to one another and to non–verbal stimuli by the mechanism of conditioned response."[31] The more theoretical the language becomes, the more the meanings of the sentences depend upon their "coherent" association with other sentences — contextual meaning — so that the supporting evidence even of observation sentences becomes progressively attenuated.

> Commonly a stimulation will trigger our verdict on a statement only because the statement is a strand in the verbal network of some elaborate theory, other strands of which are more directly conditioned to that stimulation. Most of our statements respond thus to reverberations across the fabric of intralinguistic associations, even when also directly conditioned to extralinguistic stimuli to some degree. Highly theoretical statements are statements whose connection with extralinguistic stimulation consists pretty exclusively in the reverberations across the fabric (OR, 16).

While this passage, like so many others in Quine, is beautifully expressed and very suggestive, it is nonetheless, rather vague, again relying on the concept of "linguistic association." Lacking is a detailed analysis of how the statements comprising our scientific theories are derived from experimentation, inference, theory construction, and deduction. The existence of many of the new particles in physics such as the neutron, neutrino, positron, and meson were deduced and predicted from theoretical considerations before their existence was experimentally detected. The whole theory of quarks is based on the deduction of what properties these new elementary particles must possess in order to account for the strong forces or interactions among the subnuclear hadrons (the baryons and mesons), according to the laws of symmetry.[32] What is needed is an explication of these theoretical developments based on scientific inquiry, such as we shall undertake in the next chapter, not merely their baptism as "linguistic association."

Now that we have an overall understanding of Quine's conception of how language is acquired, we can evaluate the important consequences he draws from this: namely, indeterminacy of translation, inscrutability of reference, holism, ontological relativity, and underdetermination of theories by empirical evidence. Each of these theses is implied by what might be called Quine's general position of linguistic relativity. "Linguistic relativity" was introduced by the American linguist Benjamin Whorf in a 1940 article which Quine acknowledges in a footnote in *Word and Object*.[33] Whorf, along with Edward Sapir, was one of the first to call attention to the pervasive "background influence of language" on the way we conceptualize, as well as speak about, the world. Contrasting the grammatical structures of various American Indian languages, especially the Hopi, with English, Whorf concluded that owing to their diverse grammatical structures, each imposed a different metaphysical view, or *Weltanschauung,* on their users. In Quine's terms, the ontology of these different world views is dependent upon, and therefore relative to, the background linguistic system. As Whorf states:

> the background phenomena [of a language]...are involved in all our foreground activities of talking and of reaching agreement, in all reasoning and arguing of cases....It was found that the background linguistic system (in other words, the grammar) of each language is not merely a reproducing instrument for voicing ideas but rather is itself the shaper of ideas, the program and guide for the individual's mental activity, for the analysis of impressions, for his synthesis of his mental stock in trade. Formulation of ideas is not an independent process...but is part of a particular grammar, and differs, from slightly to greatly, between different grammars.[34]

From this Whorf derives his thesis of "linguistic relativity:"

> We are thus introduced to a new principle of relativity, which holds that all observers are not led by the same physical evidence to the same picture of the universe, unless their linguistic backgrounds are similar, or can in some way be calibrated (p. 214).

Although eschewing the mentalistic terms used by Whorf, Quine's position is similar—in fact, his philosophy of language could be considered an exegesis of Whorf's theory. In his well-known example (first presented in *Word and Object* and often repeated) of the field linguist's unsuccessful efforts to precisely translate the native's use of the term 'gavagai' into English when a rabbit scurries by, Quine advances the doctrine of the "indeterminacy of translation." As in Whorf's theory, since any linguistic distinction and designation can only be made in reference to some background linguistic system, one can never be certain whether the individuative terms of one language coincide with those of another. Hence there is no empirical way of deciding, "no fact of the matter," whether the native, when he uses the term 'gavagai,' is referring to 'undetached rabbit part,' 'rabbit stage,' or 'rabbit' as an individual object *in our sense of the term*. The field linguist must depend upon various hypotheses of translation, "analytical hypotheses," but since there is no behavioral evidence (i.e., ostensive evidence) for deciding among the various hypotheses, and since one cannot appeal to a more determinate concept or meaning in the mind of the native in terms of Quine's naturalistic–behavioristic presuppositions, it is impossible to decide on the correct translation.

> If you take the total scattered portion of the spatio-temporal world that is made up of rabbits, and that which is made up of undetached rabbit parts, and that which is made up of rabbit stages, you come out with the same scattered portion of the world each of the three times. The only difference is in how you slice it. And how to slice it is what ostension or simple conditioning, however persistently repeated, cannot teach (OR, 32).

This is where Quine's naturalistic–behavioristic doctrine plays a decisive role.

Limiting himself to behavioral evidence, Quine maintains that we cannot assume that words have a more determinate meaning in the minds of the users ("the myth of the museum") than can be elicited on an occasion by behavioral criteria. But this seems to beg the question. Even if the ostensive evidence and behavioral criteria are insufficient to determine precisely whether the native means by 'gavagai,' 'undetached rabbit parts,' 'rabbit stages,' or 'entire rabbit,' why should one conclude either that there is *no such distinction in the mind of the native,* or that by further questioning *no such distinction can be elucidated*? Not infrequently in ordinary conversation we find that we are misunderstood or that a concept that seems clear to us is vague or unintelligible to someone else. Often, this is because we are mistaken about the clarity of our concepts, but sometimes the background of the other person impedes his understanding of us. In the latter case, further discussion and *reflection* by both parties can lead to clarification and agreement, but this involves more than just ostensive reference and behavioral evidence—it involves the *deliberate intent* to make our *ideas* clear and more determinate to the other person by *thinking* about what we *mean*.

Quine's elimination of the role of these familiar cognitive processes drove him to his paradoxical conclusion. He acknowledged the role of query and assent, but not the fact that this questioning is merely a behavioral manifestation of conscious processes that are inward, private, and reflective. For example, when confronted with another of Einstein's thought experiments at the Solvay conference in 1923, the "clock in the box" experiment (in which Einstein tried to show a violation of Heisenberg's energy–time uncertainty relation), Bohr was so disturbed that he was unable to sleep, passing much of the night attempting to resolve the issue. As Heinz Pagels recounts the event:

> Bohr spent a sleepless night thinking about the problem. If Einstein's reasoning was correct, quantum mechanics must fail. By morning he discovered the flaw in Einstein's reasoning. . . .Since the position of the clock is uncertain by a small amount because of the "kick" it gets when the photon escapes, so is the time it measures. Bohr showed that the thought experiment devised by Einstein did not in fact violate the uncertainty relation but confirmed it.[35]

As Pagel's account indicates, the problem's resolution depended upon processes of reasoning, not of behavioral conditioning.

From the indeterminacy of translation follows the "inscrutability of reference," because if we cannot decide what the correct translation of a sentence is, we cannot be certain of its reference.

> The indeterminacy of translation now confronting us, however, cuts across extention and intention alike. The terms "rabbit," "undetached rabbit part," and "rabbit stage" differ not only in meaning; they are true of different things. Reference itself proves behaviorally inscrutable (OR, 35).

While individuation can be determinate within a particular linguistic framework, owing to the grammatical apparatus of pronouns, identity, pluralizations, and so on, along with the logical quantifiers, we can never know whether these grammatical struc-

tures coincide with those of a radically different language. As Whorf
stated, "the activities of talking and reaching agreement" presuppose these, therefore it would be circular to *justify* an agreement in
terms of them. In words similar to Whorf's, Quine maintains that
such identifications can only be made relative to a "background
language." Given a background language,

> we can say. . .this is a rabbit and that a rabbit part, this and that the
> same rabbit, this and that different parts. *In just those words.* This
> network of terms and predicates and auxiliary devices is, in relativity
> jargon, our frame of reference, or coordinate system. Relative to *it* we
> can and do talk meaningfully and distinctly of rabbits and parts. . . .
> reference *is* nonsense except relative to a coordinate system. In this
> principle of relativity lies the resolution of our quandary (OR, 48).

Surprisingly, Quine does not limit the theses of indeterminacy
and inscrutability to radical translations, whether from a native's
tongue to our own, or from a radically different scientific theory,
such as quantum mechanics, to ordinary languages, but applies it to
individuals *within the same linguistic community,* and *even to
ourselves*!

> I have urged in defense of the behavioral philosophy of lan-
> guage. . .that the inscrutability of reference is not the inscrutability of
> a fact; there is no fact of the matter. But if there is really no fact of the
> matter, then the inscrutabilty of reference can be brought even closer
> to home than the neighbor's case; *we can apply it to ourselves* (OR,
> 47; italics added).

While acknowledging that this seems to lead to "the absurd
position that there is no difference on any terms, interlinguistic or
intralinguistic, objective or subjective, between referring to rabbits
and referring to rabbit parts or stages" (OR, 47), Quine, like Whorf,
accepts the conclusion, adopting the position of linguistic relativity.
All knowledge is framework dependent so that any ontological
claim (e.g., regarding the parts, stages, or integrity of rabbits) can
only be made relative to some linguistic framework or translation
from one language to another: "it makes no sense to say what the
objects of a theory are, beyond saying how to interpret or reinterpret that theory in another" (OR, 50).

There is a sense, of course, in which Quine's claim is true, in
that without any conceptual–linguistic framework it would be extraordinarily difficult (not impossible, because animal behavior im-

plies certain truth claims, as when a beagle is running a rabbit) to make precise truth claims, whether of a descriptive or of an ontological nature. Yet there is an air of paradox about Quine's conclusion that he himself acknowledges. After all, scientists have been able to make very subtle discriminations and difficult confirmations regarding the nature and existence of certain entities, such as the disconfirmation of phlogiston and the discovery of oxygen; the refutation of the caloric theory of heat and the fluid theory of electricity; the confirmation of Cajal's neuronal theory; Dirac's prediction and the confirmation of anti-matter; and Hideki Yukawa's theory of and the later discovery of mesons. These examples are pertinent because they do not refer merely to the existence or non-existence of accepted categories of entities; instead, their confirmation or disconfirmation raised basic questions regarding the categorical interpretation as well as the ontological status of the entities under investigation, questions which were eventually settled. Of course they were not decided by mere "ostensive reference," but by experimentation, by interacting with the object. Quine's naturalistic-behavioristic doctrine, along with his exclusive reliance on the devices of grammar and logic, and his indifference to the actual procedure and discoveries of scientists prejudices his conclusion. In his interpretation of knowledge, Quine is too much of a logician and not enough of a philosopher of science!

This is evident even in the 'gavagai' example. Recall his earlier statement that "how to slice" the world "is what ostension or simple conditioning, however persistently repeated, cannot teach." That is, without reference to a background linguistic system mere pointing and reinforcing would not result in one's being able to individuate or categorize objects unambiguously. But given the native's language, is it still reasonable to suppose that we could *never* determine whether the native meant by 'gavagai,' 'undetached rabbit part' or 'rabbit stage,' rather than the entire rabbit? Quine offers as explanation that "you come out with the same portion of the world each of the three times." While this may be true if one is limited to querying the native while a rabbit rushes by, it would not be true if one were to shoot the rabbit and examine it. Then, by grasping the head or legs instead of the whole body, one could determine whether the native used 'gavagai' in each instance or, as is more likely, that he had different words to distinguish the parts of the rabbit from the whole, as we do. Similarly, with a little ingenuity,

I think a linguist could evoke a distinction between rabbit stages and rabbit.

Furthermore, as pointed out earlier, Quine's thesis of the indeterminacy of translation seems to be contradicted by the statement that the "predicament of the indeterminacy of translation has little bearing on observation sentences" because the "equating of an observation sentence of our language to an observation sentence of another language is mostly a matter of empirical generalization; it is a matter of identity between the range of stimulations that would prompt assent to the one sentence and. . .to the other" (OR, 89). It seems to me that the situation in which both the native and the field linguist observe a rabbit running by would produce "an identity between the range of simulations" and, therefore, would "prompt assent" to "equating" their observation sentences. But if this is true, I do not understand why one would remain in doubt about their translation, especially as Quine is critical of "the tendency of Polányi, Kuhn, and the late Russell Hanson to belittle the role of evidence and to accentuate cultural relativism" (OR, 87). What could "accentuate the role of cultural relativism" more than "indeterminacy of translation" and "inscrutability of reference?"

One finds this ambivalence or tendency to have it both ways in Quine's profession of realism as well. He has maintained not only the impossibility of translating words like 'gavagai' from the native's language to ours, but believes that even "in our own case" translation is indeterminate. But if reference were indeed that indeterminate, how could one be a realist with respect to knowledge? If translations are indeterminate and reference is inscrutable *in all cases,* what is there to support a realist position? These impediments notwithstanding, Quine claims to be a realist.

> Thus adopt for now my fully realistic attitude toward electrons and muons and curved space–time, thus falling in with the current theory of the world despite knowing that it is in principle methodologically undetermined. Consider, from this realistic point of view, the totality of truths of nature, known and unknown, observable and unobservable, past and future. The point about indeterminacy of translation is that it withstands even all this truth, the whole truth about nature.[36]

Quine's thesis of the indeterminacy of translation gains further support from his holist theory which, in turn, is dependent upon what Gibson calls "the most important single statement that occurs in Quine's writings," namely, that "*a statement about the world does not always or usually have a separate fund of empirical consequences (or meaning) that it can call its own. . . .*"[37] As most sentences composing the fabric of natural languages as well as scientific theories are theoretical rather than observational, so that their meanings and evidence do not coincide with sensory stimulation, their meanings are acquired within a larger context of statements. Consequently, they are not testable in isolation but only in conjunction with large blocks of other statements, so that even the evidence for observation statements becomes inconclusive. As our theories become more complex, their empirical underpinnings become more theory-dependent, diffuse, and broadly distributed, hence more difficult to isolate and identify. Quine associates this holistic interpretation of theories with the views of Peirce and Duhem.

> If we recognize with Peirce that the meaning of a sentence turns on what would count as evidence for its truth, and if we recognize with Duhem that theoretical sentences have their evidence not as single sentences but only as larger blocks of theory, then the indeterminacy of translation of theoretical sentences is the natural conclusion (OR, 80–81).

This concept of holism can be illustrated with any scientific theory, but the special theory of relativity provides a particularly clear example. By the end of the nineteenth century a number of interrelated principles constituted part of the theoretical structure of Newtonian mechanics. They included the existence of absolute space and time underlying the notions of absolute rest and motion; the existence of an ether at rest in space for the propagation of electromagnetic waves; the revolution of the earth in its orbit around the sun at twenty miles per second; and the belief that the Galilean addition of velocities principle could be applied to all motions, electromagnetic as well as mechanical, so that the velocity of light as measured by observers in relative motion would differ. Because no mechanical motion was a reliable indication of an absolute transposition in space, Michelson and Morley (separately) tried to determine indirectly by optical means, using the interferometer, whether motion with respect to absolute space could be detected. Assuming that a light source split into two perpendicular beams but radiated through the same distances would have their velocities affected by the "ether wind" created by the motion of the

earth through the ether, Michelson and Morley expected the light
radiated across the ether wind to return slightly before that radiated
in the direction of the ether wind. Much to their astonishment,
however, they found that the two light beams returned at the same
moment. Given these apparently incontrovertible results, physicists
were faced with deciding which of the various principles presupposed by the experiment was at fault, this not being indicated by
the null results themselves. As Quine states:

> Sometimes. . .an experience implied by a theory fails to come off; and
> then, ideally, we declare the theory false. But the failure falsifies only
> a block of theory as a whole, a conjunction of many statements. The
> failure shows that one or more of those statements is false, but it does
> not show which (OR, 79).

In the case of the Michelson–Morley experiments, the negative
outcome could be attributed to the falsity of any of the stated
principles. However, when the results were interpreted in terms of
Einstein's special theory of relativity which accepted the invariant
or constant velocity of light, a block of theory was found to be at
fault, requiring the denial not only of the existence of the ether and
of absolute space and time, but also of the Galilean addition of
velocities principle, as applied to light. For if the velocity of light
was invariant, unaffected by the velocity of its source and of any
observer moving relative to it, there was no point in trying to
measure the effects of these motions on it. Thus the null results of
the Michelson–Morley experiments reverberated throughout the
theory requiring many revisions, particularly in the previous assumptions that the dimension of measuring rods, rates of clocks,
and magnitude of masses were unaffected by their velocities.

This holist interpretation of theories had a marked influence on
Kuhn's and Feyerabend's contention that theories which undergo
radical change are "incommensurable." For if the meanings of the
key terms comprising the transformed theory do not belong to the
terms individually, but depend upon the interrelation of meanings
within the theory, then, as the theory changes, so do the meanings
of the key terms. Furthermore, not only is the content of the
theories radically different, what they refer to is entirely different,
because terms having completely different meanings (if the theories are sufficiently different) *cannot refer to the same physical
elements or states.* Such theories do not undergo rational transi-

tions, they suffer discontinuous changes and, therefore, are incommensurable. This question of whether theory change can be construed as rational, a central and acute problem in current discussions in the philosophy of science, will be considered further in the next chapter. It was referred to here because of the influence of Quine's holist theory on the problem.

As indicated previously, one component of the holist theory consists of Peirce's verifiability theory of *meaning*: that whatever counts as evidence for the truth of an assertion constitutes its meaning. As this was a tenet the positivists subsequently rejected, one is surprised to find Quine still endorsing it. Because the evidence and the meaning of observation sentences tend to converge, one *can* maintain the verifiability criterion of meaning as regards these statements; however, since the meanings of theoretical terms and statements depend upon contextual definitions involving much more abstract and theoretical concepts, they are not subject to direct tests, as Quine's holist theory correctly maintains. In what sense, then, can one say that their meaning consists in what would verify them? In no sense at all, as the positivists came to realize. The very attempt to confirm a statement presupposes that one understands it. And since the evidence for the truth of a statement is often unknown prior to the testing procedure, the meaning of the statement cannot consist of the evidence that would verify it. The *truth* of the assertion depends upon the evidence, and so does the *refinement* of meaning during the verifying procedure, but the evidence does not constitute the *original* meaning, though it can become part of the meaning *later.*

A clear illustration of this appeared in a recent newspaper report of the theory that the extinction of dinosaurs was caused by the collision of a comet with the earth which produced debris that blackened the sky for years, the consequent darkness and cold killing many plant and animal species.[38] When first formulated five years ago by Nobel physicist Luis Alvarez and his geologist son Walter, it was not taken seriously, the accepted view being that gradual climactic changes over millions of years caused the extinction. However, when father and son began examining the thin layer of earth that was deposited when the dinosaurs perished, they found that it contained abnormally high concentrations of iridium, a slowly decaying element found in abundance in comets but not in the earth's crust. It was this unexpected evidence, *not contained in the original theory,* that has given credibility to it.

Because the theorist possesses a stock of concepts and meanings which are conventionally associated with words, he or she can express new hypotheses and conjectured theories prior to knowing what evidence would confirm them, as is true often of new scientific theories. This is what distinguishes theoretical statements from observation statements; the latter wear their empirical evidence on their sleeves, whereas the former share their evidence (acquired in ways involving much more than stimulation of sense receptors) with other sentences. For example, there is considerable disagreement whether psychoanalytic theories can be empirically tested, but hardly anyone would maintain today that they are, for that reason, meaningless. Each psychoanalytic school provides its own framework of interpretation, with its particular definition of basic concepts which are intelligible to its users, even though these may not be testable in a reliable or objective sense. The fact that words do have a *determinate* conventional meaning as understood by individuals, and so can be used in different contexts to construct new theories, is what makes the formulation of theories prior to their testing possible.

Each component of Quine's position, indeterminacy of translation, inscrutability of reference, linguistic relativity, and holism is implied in the last thesis, underdetermination of theories by empirical evidence, or its corollary (not stated by Quine), the overdetermination of empirical evidence by theories. As succinctly summarized by Roger Gibson:

> Essentially, the doctrine is that physical theory (i.e., our theory about the world), transcends our observations of the world, and that different, competing, physical theories can, therefore, be developed from the same set of observations. In a word, physical theory is *underdetermined* by the observations that support it, thereby engendering a fair amount of empirical slack in our beliefs about the world. . . .[39]

According to Quine, even if we possessed all true observation sentences this would not mean that only one theoretical interpretation of these sentences was possible, so any interpretation must extend beyond the empirical evidence. Furthermore, because "the typical statement about bodies has no fund of experiential implications it can call its own" (OR, 79), it is not equivalent to any set of observation sentences, as Carnap once believed. If different theoretical interpretations of the same set of observation statements are possible, it follows that all observation statements are overdetermined by theory or, as Quine puts it, all theories are underdetermined by empirical evidence.

> Theory can still vary though all possible observations be fixed. Physical theories can be at odds with each other and yet compatible with all possible data even in the broadest sense. In a word, they can be logically incompatible and empirically equivalent. This is a point on which I expect wide agreement, if only because the observational criteria of theoretical terms are commonly so flexible and fragmentary.[40]

In this case, Quine's conclusion seems to be influenced by the actual conditions in science in agreement with Einstein's thesis that theoretical concepts are not deduced from empirical data, but are "free creations" of the mind, thus transcending the data, and following from Carnap's failure to translate the theoretical statements of science into observation sentences. Given that theoretical explanations are not deducible from the empirical evidence and have no unique empirical consequences of their own, they remain underdetermined as regards their observational support. This was the meaning of the net simile in Wittgenstein's *Tractatus*. Theories are like nets that can be stretched over the empirical data without catching everything inherent in them and, therefore, differ in what they contain: "The different nets correspond to different systems for describing the world" (6.341). The net simile not only depicts what is essential in Quine's position, it conforms to Popper's notion of theory falsification as leading to a progressive approximation of truth. However, Wittgenstein goes further than both Quine and Popper, who claim to be realists, when he adds that "the possibility of describing" something "with a net of a given form tells us *nothing* about" it (6.342). Here, as usual, Wittgenstein would seem to be wrong. It is only because a net (or theory) has something in common with the world, that it contains concepts and a form designating some features of the world, that it can be used to describe it and, therefore, *does* tell us something about those features, as I shall argue more fully in the next chapter.

While Quine's notion of the underdetermination of theories by empirical evidence is well founded (though he later had misgivings about the intelligibility of the thesis,[41] the reason for its validity does not seem to be fully understood by Quine. In my opinion, the reason all significant theories are underdetermined by empirical evidence, in the sense that they cannot be deduced from the evidence and that their empirical consequences are not fully specifiable, is because *the description of nature can never be exhausted in observation sentences*. The domain of observable phenomena is exceedingly restricted, dependent upon our limited sensory capacities which do not reveal the extensive dimensions of physical reality reaching inward to the nucleus of the atom and stretching outward to the extremity of the universe. The necessity of probing these domains to explain our ordinary observations has given rise to experimental science, yet Quine never acknowledges the crucial role of experimentation in the discovery and explanation of phenomena. He is so preoccupied, as were the positivists, with observation sentences that he seems unaware of the crucial function of experimentation in probing nature to obtain new data for the construction and verification of theories.

Moreover, it is due to these experimental inquiries, not mere observations, that we discover that the presuppositions embedded in our common sense beliefs and scientific theories are inadequate when applied to less familiar dimensions or levels of reality. Our theories do presuppose much more than is explicitly implied by their known observable consequences. The Michelson–Morley experiments uncovered the unsuspected constant velocity of light; Rutherford's early experiments on gold foil with alpha particles disclosed the existence of dense particles surrounded by empty space; and Planck's pursuit of the ultraviolet catastrophe resulted in his discovery of an irreducible, discontinuous quantum of interaction. The remarkable discoveries in particle physics have coincided with the development of cyclotrons and linear accelerators, as well as the cloud chamber, so that we are now aware that what can be discovered is a function of the velocity or energy imparted by the apparatus to the accelerated particles used to collide with the target. Although Quine has justly rejected the notion of "a first philosophy, higher than physics," he still views science as a layer of theoretical statements terminating in the firmament of observation statements, rather than as the continuous exploration of deeper domains of physical reality that seem inexhaustible in their novelty and dimensions, and therefore require continuous revisions in the interpretation of observations, as well as in the theoretical framework.

In conclusion, the purpose of this chapter has been to describe some of the dominant influences and tendencies in twentieth century philosophy, culminating in the system of Quine. While I have identified points of disagreement with Quine, it remains to be shown whether a more adequate treatment of these problems can be offered, one more attuned to recent developments in the philosophy of science, especially the image of science as a dynamic method of inquiry used to explore unknown recesses of the universe, with consequences that have tranformed the conditions of human existence as well as our ways of thinking (or "talking") about the world. In the following chapter, I shall try to justify a conception of reference and meaning based on a more adequate understanding of scientific explanation, and a more realistic interpretation of scientific theories.

Notes to Chapter III

1. Wilfrid Sellars, *Science, Perception and Reality* (New York: Humanities Press, 1963), p. 60.
2. Cf. Harold I. Brown, *Perception, Theory and Commitment* (Chicago: University of Chicago Press, 1979), p. 30.
3. Cf. O. Neurath, R. Carnap, and C. Morris (eds.), *International Encyclopedia of Unified Sciences,* Vol. I (Chicago: University of Chicago Press, 1955), pp. 29–39.
4. Cf. John Dewey, *Quest for Certainty* (New York: G. P. Putnam's Sons, 1929), chs. V, VIII.
5. Cf. A. Einstein, "Autobiographical Notes," in *The Library of Living Philosophers,* Vol. VII, ed. by Paul A. Schilpp (Evanston: Library of Living Philosophers, 1949), p. 21. Also, Percy Bridgman, *The Logic of Modern Physics* (New York: Macmillan Co., 1927).
6. Ludwig Wittgenstein, *Philosophical Investigations,* trans. by G. E. M. Anscombe (Oxford: Basil Blackwell, 1953), Part I, sec. 122. Subsequent references to this work are indicated by the Part and the section.
7. Bertrand Russell, *The Problems of Philosophy* (London: Oxford University Press, 1912), ch. V. It was in this book that Russell also introduced the term "sense data," p. 12.
8. Saul Kripke, *Naming and Necessity* (Cambridge: Harvard University Press, 1972), p. 64.
9. Cf. Richard H. Schlagel, "Contra Wittgenstein," in *Philosophy and Phenomenological Research,* June, 1974, pp. 539–550.
10. Cf. Karl Popper, *Objective Knowledge* (Oxford: Clarendon Press, 1972), p. 57.
11. Cf. Norwood Hanson,. *Patterns of Discovery* (Cambridge: Cambridge University Press, 1961), p. 19.
12. Cf. Thomas S. Kuhn, *The Structure of Scientific Revolutions,* 2nd ed. enlarged (Chicago: University of Chicago Press, 1970), p. 150.
13. Paul Feyerabend, *Against Method* (London: NLB, 1975), p. 23.
14. Cf. Paul Feyerabend, *Science in a Free Society* (London: NLB, 1978), p. 39.
15. W. V. Quine, *Ontological Relativity and Other Essays* (New York: Columbia University Press, 1969), p. 82. Subsequent references to this work will be indicated in the text by OR and the page references.
16. W. V. Quine, "Two Dogmas of Empiricism," in *From a Logical Point of View* (Cambridge: Harvard University Press, 1953), p. 20.

17. Roger F. Gibson, *The Philosophy of W. V. Quine* (Tampa: University Presses of Florida, 1982), pp. xviii–xix.
18. W. V. Quine, "The Nature of Natural Knowledge," in *Mind and Language,* ed. by Samuel Guttenplan (Oxford: Oxford University Press, 1975), p. 78.
19. Roger Gibson, *op. cit.,* p. xx.
20. W. V. Quine, "Mind and Verbal Dispositions," in *Mind and Language, op. cit.,* p. 83. The immediately succeeding references are to this work unless otherwise indicated.
21. Whereas Gibson believes that Quine's "behaviorism. . .countenances introspection and mentalistic terms" (*op. cit.,* p. 196), I think, as the earlier quotations clearly indicate, this was not true when he wrote OR.
22. W. V. Quine, "Four Hot Questions in Philosophy," a review of P. F. Strawson's *Skepticism and Naturalism, The New York Review of Books,* February 14, 1985, p. 32.
23. Cf. Noam Chomsky, *Language and Mind,* enlarged ed. (New York: Harcourt Brace Jovanovich, 1968), p. 102.
24. Cf. *ibid.,* p. 14.
25. W. V. Quine, "Philosophical Progress in Language Theory," in *Metaphilosophy,* 1970, No. 1, p. 5. Quoted from Roger Gibson, *op. cit.,* p. 203.
26. A personal communication from Quine to Roger Gibson, *op. cit.,* p. 204.
27. Derek Bickerton, "Creole Language," in *Scientific American,* July, 1983, p. 121. Also see his *Dynamics of a Creole System* (Cambridge: Cambridge University Press, 1975).
28. Paul Churchland, *Scientific Realism and the Plasticity of Mind* (Cambridge: Cambridge University Press, 1979), pp. 24–25.
29. W. V. Quine, *Word and Object* (Cambridge: Harvard University Press, 1960), p. 9.
30. Cf. W. V. Quine, *The Roots of Reference* (La Salle: Open Court Publishers, 1974). Also, Roger Gibson, *op. cit.,* pp. 16–28; 41–62.
31. W. V. Quine, *Word and Object, op. cit.,* p. 11.
32. Cf. James S. Trefil, *From Atoms to Quarks* (New York: Charles Scribner's Sons, 1980), chs. IX, X; also Harold Fritzsch, *Quarks* (New York: Basic Books, 1983), ch. 14.
33. W. V. Quine, *Word and Object, op. cit.,* p. 77.
34. Benjamin Lee Whorf, *Language, Thought, and Reality,* ed. by John B. Carroll (New York: M.I.T. Press and John Wiley & Sons, 1956), pp. 212–213. Brackets added.
35. Heinz R. Pagels, *The Cosmic Code,* (New York: Simon & Schuster, 1982), p. 97.
36. D. Davidson and J. Hintikka (eds.), *Words and Objections* (Dordrecht-Holland: D. Reidel Publishing Co., 1969), p. 303.
37. Roger Gibson, *op. cit.,* p. 3.
38. Cf. Philip J. Hilts, "Comet-Extinction Link Gains Support," in *The Washington Post,* April 20, 1984, pp. A1, A12.
39. Roger Gibson, *op. cit.,* p. 84.
40. W. V. Quine, "On the Reasons for Indeterminacy of Translation," in *The Journal of Philosophy,* No. 67, 1970, pp. 179–183. Quoted from Roger Gibson, *op. cit.,* p. 84.
41. Cf. Roger Gibson, *op. cit.,* p. 85.

CHAPTER IV

THE RATIONALITY OF SCIENCE[1]

IN previous chapters it has been maintained that an adequate conception of the foundation, structure, and limits of knowledge depends upon a correct assessment of the knowledge we possess. Since at this time in history, the various sciences provide us with the most reliable knowledge, it is to science that we must look for our model of understanding knowledge claims. Yet science (as the recent revolution in the philosophy of science following the collapse of logical positivism has shown) is not a fixed, static discipline any more than are art or literature, but an evolving methodology and technology, with a continuously developing theoretical content. Thus we can identify much in Aristotle's system that is scientific, though his methodology lacked experimentation and the search for mathematical correlations (as opposed to ratios alone), because he *was* concerned with explaining natural phenomena, and his emphasis on observation was a necessary stage in the evolution of science.[2]

If science is a dynamically evolving discipline, any critical evaluation of its attainments will necessarily depend upon its level of development at the time. This is clearly evident in David Hume's celebrated eighteenth century critique of causality and induction,

although it has been overlooked by prominent contemporary philosophers who appear to accept the finality of Hume's assessment.
Peter Strawson, for example, states: "If. . .there is a problem of
induction, and. . .Hume posed it, it must be added that he solved
it. . ."[3] Quine is even more emphatic: "On the doctrinal side [i.e., as
regards the truth of the doctrine], I do not see that we are farther
along today than where Hume left us. The Humean predicament is
the human predicament."[4] These claims notwithstanding, I think it
can be shown quite convincingly that Hume's critique of causality
and induction, like his conception of the origin and structure of
knowledge on which it so fatally depends, though then a stunning
insight, *betrays the limited understanding, at the time, of knowledge in general and of scientific knowledge in particular.* When
examined within the context of the explanatory limitations of
eighteenth century science, the rationale for Hume's position becomes eminently clear, along with the resolution of the problem.

As Hume's analysis of knowledge into atomic impressions and
their replication as faint copies or ideas is regarded as totally inadequate today, *so his notion that scientific inferences are limited to
the succession of ideas, or even to the observable qualities of
objects, is decidedly contrary to present scientific attainments.*
Recent analytic philosophers, especially the positivists, were prevented from seeing the solution to Hume's problem by their anti-
historical approach. If, as Reichenbach claimed, the "justification" of
a view is to be kept entirely separate from the context of its
"discovery" or development, then one is precluded from considering the historically derived presuppositions that contributed to the
generation of the problem, presuppositions that may appear less
plausible from a later perspective. Yet understanding the historical
context within which a traditional problem was conceived can lead
to a clarification of the problem, along with its resolution. However,
the realization of this required a revolutionary change of view, one
that is generally accepted now by philosophers of science.

Accordingly, a way to better understand contemporary scientific
achievements is to see what was behind Hume's skeptical assessment of science as it existed in the eighteenth century. Hume was
aware that no experience or knowledge of the world acquired in the
past would have the least significance for the future but for two
beliefs: (1) that what we encounter in the future will resemble what
we have previously experienced (otherwise our past experience
would be irrelevant), and (2) that whatever "hidden mechanisms"

or "secret causes" produced these similarities will remain the same. Hume never questions the first belief, presumably because it is so obvious and continuously confirmable, directing his argument against the second. Even when we can identify objects and events as similar to those experienced in the past, we have no *rational* grounds for believing that whatever underlying mechanisms or causes that produced those objects and events have either remained the same or continued to be operative. In other words, recurring similarities do not constitute evidence for the identity or continuation of what produces them, hence no expectations can be based on them. As our experience is limited solely to the successive appearances and resemblances of objects, we do not *experience* their underlying causes; therefore, without some additional *knowledge* of a *necessary* connection between the appearances and the causes, the similiarities themselves are not evidence of any continuation or identity in what produces them. As Hume states in the *Enquiries*:

> In vain do you pretend to have learned the nature of bodies from your past experience. Their secret nature, and consequently all their effects and influence, may change, without any change in their sensible qualities. This happens sometimes, and with regard to some objects: Why may it not happen always, and with regard to all objects? What logic, what process of argument secures you against this presupposition?[5]

It must be emphasized, because many philosophers have misread this, that Hume is *not* denying the *existence* of "secret natures" and "powers," but that we have any basis for claiming to have knowledge of them!

Having disclaimed any evidence of a connection between the sensible qualities of objects and their "secret natures" or "powers," Hume's ontology comes to a strange impressionistic succession of phenomenal qualities which do not reveal any natural causes or productive processes. All that is productive or causal is hidden: volcanos, hurricanes, forest fires, tidal waves, or epidemics are not manifestations of discoverable underlying processes and causes, but cinematic sequences whose mere succession creates the illusion of causal production. Naturally, if there were *no evidence whatsoever* of the dependence of the observable qualities of objects and their effects on the inherent natures of the objects, then the qualities could be viewed as entirely detached from any inner causes, and therefore could remain unchanged while the causes changed.

But were this the case, one wonders why we ever would have been deluded into thinking there was a dependent connection in the first place!

Even concerning directly observable phenomena, such as volcanos, tidal waves, avalanches, or forest fires, is it plausible to suppose that we do not observe the causal agents producing the effects, only the succession of appearances? While it is true that we do not perceive all the minute mechanisms or interactions involved, still, when we witness a forest fire, for example, do we not see the flames devour the underbrush and trees? As Harré and Madden state in their excellent critique of Hume,

> the argument which gives ontological status to causal powers in general also applies to the everyday situations where we talk about the waves eating away the shore, the axe splitting the wood, and the avalanche destroying the countryside, and hence that it is possible for one to perceive processes in which the agency of causal powers is manifested directly.[6]

In any case, today we no longer are limited to normal perception for our evidence of causal influences. Our capacity for witnessing the more intrinsic states and interactions of nature has been dramatically extended by the use of highly refined telescopes, microscopes, X-ray spectroscopy, and computer graphics. Tissues and cellular structures that previously were indiscernable are now observed under the microscope with the proper staining preparation. The extraordinary cellular development of a human embryo can be currently witnessed thanks to recently improved photographic techniques. In the past several years advances in computer graphics have led to the production of three-dimensional representations of such molecular structures as the common cold virus and DNA, the latter's atomic symmetry resembling that of the rose window in the National Cathedral in Washington. A newly invented microscope capable of magnifying an object 300 million times presently allows scientists to take pictures of assemblies of individual atoms composing the surface of ordinary objects. By accelerating and colliding particles using the cyclotron, synchrotron, linear accelerator, and the new Tevatron at Fermilab, physicists, using the Wilson cloud and bubble chambers, have been able to indirectly observe, photograph, and make computer simulated images of the intricate collisions, annihilations, and creations of subatomic particles. Hume, of course, cannot be held responsible for failing to foresee these developments, but their occurrence indicates how mistaken he was in proscribing observation of "insensible" causes or powers.

In fairness to Hume, there is a sense in which his view could be considered partially correct, in that human beings (or as he would say, "human nature") tend to confuse familiarity with understanding.

> The generality of mankind never find any difficulty in accounting for the more common and familiar operations of nature—such as the descent of heavy bodies, the growth of plants, the generation of animals, or the nourishment of bodies by food: But suppose that, in all these cases, they perceive the very force or energy of the cause, by which it is connected with its effect, and is for ever infallible in its operation (E, p. 69).

Granting this, there is still a radical difference between the insight that people deceive themselves in thinking they understand familiar phenomena, and the claim that they are deluded in believing that the ordinary occurrences in the world are actual manifestations of productive causes—in contrast to detached, free-floating, discrete qualities. His bizarre disconnection of the observable effects of phenomena from any underlying causes acquires its ultimate rationale in Hume's empiricist epistemology, because if what we experience are *mental* impressions or ideas (rather than the actual world), then they *would be* completely dissimilar from any *physical* agents or powers.

To justify our belief that the experienced similarities in nature are indeed reliable indicators of the same productive causes or agencies, we would need to have, according to Hume, knowledge of a "necessary connection" between the observable similarities and their underlying causes. If we knew that appearances were *necessarily* connected with certain causes, then, given either the appearances or the causes, we would be justified in expecting the occurrence of the other. Similarly, though recurrent cause and effect sequences are contiguous in space, successive in time, and constantly conjoined, without any knowledge of a *necessary* connection between the cause-effect sequences, we would not be justified in inferring the one, given the other. While these three relations are *necessary* conditions for establishing cause and effect connections, they are not *sufficient*. Because rationalists like Descartes had assumed that "there must...be as much reality in the...cause as in the effect,"[7] Hume is concerned to deny such conceit: "The mind can never possibly find the effect in the supposed cause, by the most accurate scrutiny and examination" (E, p. 29).

Not only is it impossible to find the effect *in* the cause, we cannot establish any binding tie *between* the cause and the effect, because to do so we would have to be able to "intuit" or "demonstrate" an "invariable" connection between them. Analogously, to justify the belief that objects resembling those in the past will react in the same manner, we would have to acquire either an intuitive or a demonstrative knowledge of how the natures or powers of objects produce their appearances and behaviors—a knowledge completely beyond us, according to Hume.

> No object ever discovers [i.e., discloses], by the qualities which appear to the senses, either the causes which produced it, or the effects which will arise from it; nor can our reason, unassisted by experience, ever draw any inference concerning real existence and matter of fact (E, p. 27; brackets added).

With the exception of some extreme rationalists, I do not know of anyone who maintained that reason, "unassisted by experience," could infer the inherent natures and underlying causes of phenomena. The question is, *given experience,* including experimentation, can reason infer the underlying properties and causes of things, at least to some degree? This, too, Hume denies. Not only are causal powers not revealed in the observable qualities of objects, we cannot *infer* such powers from these qualities: "Should it be said. . .we *infer* a connection between the sensible qualities and the secret powers; this, I must confess, seems the same difficulty. . . .The question still recurs, on what process of argument this *inference* is founded?" (E, pp. 36–37). In actual practice today, the "process of argument" is a hypothetico–deductive one based on experimental evidence.

Along with a lack of observational and inferential evidence to justify a connection between inherent causal properties and their effects, Hume cites another criterion at least as important: contradiction. If there were discoverable "necessary connections" between the secret natures of objects and their manifest properties and effects, it would be contradictory to deny such connections. As he says in the *Treatise*: "Such a connexion wou'd amount to a demonstration, and wou'd imply the absolute impossibility for the one object not to follow, or to be conceiv'd not to follow upon the other: Which kind of connection has already been rejected in all cases."[8] Or, as he says in another passage, "Such an inference wou'd amount to knowledge, and wou'd imply the absolute contradiction and impossibility of conceiving any thing different" (T, p. 87). However, as it appears that we can always deny the truth or conceive the contrary of any matter of fact or empirical assertion, no necessary connections can be attributed between the inner natures or powers of objects and their sensible qualities and effects, or to the causal relations among objects; that is, we can always deny, without apparent contradiction, that the sun will rise tomorrow, that fire will burn paper, or that bread will nourish the human body.

Again, there is an obvious sense in which Hume's view is correct and a deeper sense in which it is not. Obviously, the semantic relation between the terms and phrases 'the sun' and 'will rise tomorrow,' or 'fire' and 'will burn paper,' or 'bread' and 'will nourish the human body' is not as self-evident as that between 'red' and 'being colored,' or 'iron' and 'being a metal,' or 'father' and 'having produced a child.' While a denial of the latter expressions would clearly be contradictory, this is not as apparent regarding the former. But this difference depends upon cultural conditioning as manifested in linguistic conventions which, in turn, reflect the level of knowledge at the time. Statements that at one time were believed to be necessary or even analytic, such as Aristotle's claim that "material objects naturally fall to the center of the universe" or Newton's assertion that "the flowing of absolute time is not liable to any change," based on the meanings or self-evident connection of the concepts, later have been falsified. On the other hand, assertions which were merely contingent at an earlier stage in science later can appear necessary, as Aristotle's rejection of the vacuum to account for the difference in motion of objects subjected to the same impact became the inviolable medieval principle "nature abhors a vacuum;" and though Newton denied that gravity was "essential to bodies," physicists later considered gravitational attraction an inherent property of mass. Thus, whether statements appear to be necessary can vary from generation to generation and from individual to individual.

As I shall argue more fully later, within the context of Newton's theory of mechanics (comprising the universal law of gravitation and the laws of dynamics), it *would be contradictory* to maintain that, under exactly the same physical conditions and without any added external influence, the earth would cease to rotate on its axis and the sun fail to maintain its orbital position, thereby preventing the apparent rising of the sun. Newton's laws imply that this contrary occurrence could not take place without some intervening cause. Similarly, given what we know about the radiant energy of fire and the chemical composition of paper, or the nutritional properties of bread and the digestive requirements of the human body, *it would be contradictory* to deny that fire will burn paper or that bread will nourish the human organism. What we now know about these objects would ensure their having these effects under the usual conditions, though to deny this may not be blatantly contradictory. Although the semantic relations between these empirical concepts have not achieved the conventional status of being analytic or self–evidently true, their denial, *given what we know today of their natures,* would be so paradoxical as to be contradictory to the specialist in that field. What this means is that theoretical explanations have necessary implications or consequences based on the meanings of the fundamental concepts and their contextual interrelations which, though obvious to the specialist, may not be evident or have the appearance of being analytic or self–evident within the context of ordinary language. Even the sentence, "an unsupported physical object in the gravitational field of the earth will not fall," is not self–contradictory in terms of the sentence's meaning, but would be self–refuting to any physicist. As Harré and Madden state:

> It does not follow. . .that a general theory which explains all sorts of particular cases of causal efficacy, such as gravitational theory, is necessarily true in the sense that its meaning entails its truth. Gravitational theory is not necessarily true in that sense though its necessity in *this world* derives from the fact that it is sufficiently fundamental to be in part definitive of the nature of this world. . .the point is that given some general theory specifying the fundamental causal powers and thereby laying down the general lineaments of a world, the necessity of certain effects can be inferred (p. 15).

It is significant that this will remain true despite the fact that the theoretical framework within which these necessary implications occur proves to be false. We no longer consider Aristotle's cosmology to be true, but within that system it necessarily follows that "a terrestrial object tends to fall towards the center of the universe" while a "celestial body moves with an eternal circular motion," these consequences being implied in the very meaning of "terrestrial object" and "celestial body." Similarly, given what we now believe about solar bodies based on empirical discoveries (e.g., their mass, inertia, and gravity), and what we know about the nature of fire and paper or bread and the human digestive tract, *it would be inconsistent* to maintain that certain predictable effects would not take place under conditions similar to those that prevailed when the discovery and explanation of these effects occurred. *Given experimentally ascertained regularities and interdependencies which are explained in terms of the discovered natures and powers of things, the necessity of certain properties, effects, and behaviors of bodies, while not analytically connected in any obvious sense, is inferred by the scientist, therefore the denial of the necessity would be inconsistent.* These *necessary* empirical consequences, following from accepted scientific theories, enable scientists to make the deductions essential to scientific prediction and discovery.

Following tradition, however, in which necessity was chained to *a prioricity* and contingency riveted to *a posterioricity,* Hume denies that there can be *necessary* connections among empirical statements. Yet for there to be predictable connections among matters of fact, he nonetheless claims that they would have to exhibit the *same kind* of necessity found in mathematics and logic, self-evidence or deductive implication. Maintaining an absolute cleavage between "relations of ideas" and "matter of fact" knowledge, and holding that the former consist of "intuitive" or "demonstrative relations" among certain ideas that can be established by reason independently of experience, Hume, surprisingly, demands that the network of empirical, factual, or scientific relations display the same *kind* and *degree* of analytical or implicative connectedness. Then, not finding such binding connections among empirical relations or ideas, he concludes that the necessity cannot be objective but must be derivable from the mind itself.

> Thus as the necessity, which makes two times two equal to four, or three angles of a triangle equal to two right ones, lies only in the act of understanding, by which we consider and compare these ideas; in like manner the necessity or power, which unites causes and effects, lies in the determination of the mind to pass from the one to the other... 'Tis here that the real power of causes is plac'd, along with their connection and necessity (T, p. 166).

In this important passage, Hume indicates that as far as mathematical truths are concerned, the necessity "lies...in the act of understanding by which we consider and compare these ideas," not in some conditioned, psychological "determination of the mind." As regards empirical knowledge, however, given his epistemological assumptions and the limitations of scientific knowledge at the time, he did not realize that a weaker kind of necessity inherent in the fundamental concepts, principles, and laws of science, abetted by experimentation, could exist. Instead, he concluded that the necessity was based on a conditioned psychological compulsion that "determines the mind" to make causal connections and inductive inferences. Given the resemblances among recurring objects and cause–effect sequences, human nature is such that the mind acquires the "habit" or "custom" (which implies that at least human nature is subject to causal effects) of expecting these recurrences to obtain with the same consequences. It is this *subjective propensity* based on recurring resemblances that constitutes the original feeling or *impression* from which the *idea* of necessity arises. Claiming that justification for these recurrences could not be found in nature, he attributed them to the mind.

> The idea of necessity arises from some impression. There is no impression convey'd by our senses, which can give rise to that idea. It must, therefore, be deriv'd from some internal impression, or impression of reflection. There is no internal impression...but that propensity, which custom produces, to pass from an object to the idea of its usual attendant. This therefore is the essence of necessity. Upon the whole, necessity is something, that exists in the mind, not in objects...(T, p. 165).

This conclusion meshes with Hume's epistemological position that derives all ideas (excluding an intervening shade of color) from faint copies of prior impressions. He had no conception of ideas in the Einsteinian sense of "free creations of the human mind," or of the modern notion of language and theories as consisting of inferred concepts and theoretical constructs, since, for Hume, even complex ideas are formed principally by association. As impressions are discrete, disconnected sensory elements, so are their faint copies or ideas, reinforcing the position that such cognitive elements are so completely independent that no necessary connection can be intuited or demonstrated among them, and hence no denial of their conjunction or recurrence can be contradictory. This empiricist epistemology is what underlies Hume's analysis of causality and induction. As described by Harré and Madden, the Humean event, the impression, basic to his epistemology, is. . .instantaneous in nature, punctiform and elementary, and from this characterisation follows its atomicity, its lack of internal connections with anything else. The atomicity of events in turn ensures the sequential independence of properties. . . .Consequently, from the independence of co-existing properties it follows *in principle* for the Humean that there can be no basis for any principle of the continuity of identity (p. 110).

If we could intuit or demonstrate necessary connections among phenomena, such inferences "wou'd imply the absolute contradiction and impossibility of conceiving any thing different. But as all distinct ideas are separable, 'tis evident there can be no impossibility of that kind" (T, p. 87). Thus Hume's impressionistic ontology can more accurately be described as pointilistic. But this conception, as we now know, is completely mistaken. It was pointed out in the previous chapter that theoretical statements cannot be reduced to observation statements, and that even sensory stimuli, to be communicated, must be reported by means of some conceptual-linguistic framework. Furthermore, as indicated in Chapters I and II, the human organism is not a passive recipient of discrete sense impressions because it processes all external stimuli by inherent neurological structures. Any sensory content, as experienced even by the infant, is preorganized to some extent. Hume's notion of atomic sensory impressions producing ideational copies is an erroneous conception of knowledge based on common empiricist misconceptions of the seventeenth and eighteenth centuries.

Consistent with his criterion that the contrast between ideas and impressions depends on nothing more than their vividness, Hume attributes any difference between opinion and belief, or degrees of belief, to relative force or liveliness: "an opinion or belief is *nothing but a strong and lively idea deriv'd from a present impression related to it*. . ." (T, p. 105). There are no grounds for a *rational* justification of beliefs based on evidence and inference; only the impact of ideas constitutes belief, the stronger the impact the greater the belief. It naturally follows from these epistemic assumptions that we cannot have any knowledge of nature as such, or any *reasonable* justification for believing in the uniformity of nature. In contrast to mathematical knowledge or deductive implications among ideas, which *do* exhibit necessary connections but do not constitute knowledge of nature, whatever apparent systematic con-

nections occur among our empirical ideas or matter of fact knowledge derive from subjective sensations or feelings, and consequently are irrational. (One is reminded of Paul Feyerabend.)

> Thus all probable reasoning is nothing but a species of sensation. 'Tis
> not solely in poetry and music, we must follow our taste and senti-
> ment, but likewise in philosophy [including natural philosophy or
> science]. When I am convinc'd of any principle, 'tis only an idea,
> *which strikes more strongly upon me.* When I give the preference to
> one set of arguments above another, I do nothing but decide from my
> *feeling* concerning the superiority of their influence. *Objects have no
> discoverable connection together;* nor is it from any other principle
> but custom operating upon the imagination, that we can draw any
> inference from the appearance of one to the existence of another
> (T, p. 103; brackets and italics added).

But *are* the evidence and reasoning supporting our scientific
theories based solely on psychological conditioning and subjective
feelings, with the result that the extensive body of exact knowledge
acquired in physics, chemistry, biology, and astronomy since
Hume's day has no relevance whatsoever for our expectations and
understanding of nature? Hume, of course, acknowledged that our
ordinary behavior and beliefs belie his skeptical conclusions, but he
challenged philosophers to provide a *rational justification* for our
confidence in induction, a challenge which I shall now take up.

Hume is largely correct in maintaining that our everyday experience of the world does not directly disclose the underlying causal
powers (i.e., the unobservable infrastructures, forces, etc.) on
which the manifest qualities and occurrences of the world depend
(although as I pointed out earlier, our ability to observe these
causes has been greatly enhanced by the use of modern instruments). When we observe any ordinary object or scene before us,
for example, we do not see the radiant energy that illuminates the
visual aspects of the world, nor do we observe this radiation being
refracted on our retinas and transmitted as chemical–electrical discharges through the optic nerves to the visual centers of the brain.
None of this is observed by us. What we experience is the outcome
of this tremendously complex but necessary process: an illuminated
field of perception consisting of various objects with their inherent
physical properties and qualities arrayed in space. Similarly, when
we feel an object to be cold, hard, and metallic we do not perceive
the atomic–molecular structure of the object that accounts for

these particular tactual properties, in contrast to observing a fire or a liquid. We know that ordinary water has a precise specific gravity and a definite freezing and boiling point, but initially we knew nothing of the molecular structure, chemical bonds, and electromagnetic weak forces or exchange of virtual photons which we now believe account for these properties. Even in saying this much, however, we betray the fact that we are not limited to mental impressions and ideas in our experience of the world. Though its occurrence may be inexplicable, our experience is not of sensory images but of an objective world of physical objects with their properties and effects, however conditioned or dependent this experience may be on the human organism.

However, while these limitations pertain to initial observations, our knowledge is not restricted, as Hume repeatedly claims, to the external appearances of things. In a way that no eighteenth century philosopher—or scientist, for that matter—could anticipate, our investigations of nature, supplemented by sophisticated instruments of observation, controlled experiments utilizing ever more elaborate apparatus, physical and chemical analysis, mathematical extrapolation and theorizing, have augmented our knowledge tremendously. Hume not only repeatedly affirms that explanations which are commonplace today were not accessible then, he asserts that they could *never* be attained, the phenomena being "incomprehensible."

The impact of billiard balls, the production of sound by a vibrating string, the occurrence of "palsy," the control our nerves exercise over the muscles of the body, and the digestion of bread not only were inexplicable at the time, any explanation was intrinsically "inconceivable," on Hume's view, which in fact it was in terms of his epistemic assumptions! As he claims regarding the cause of movement by physical impact: "We are ignorant...of the manner in which bodies operate on each other: Their force or energy is entirely incomprehensible" (E, p. 72). Hume, of course, had no idea of the energy of motion being transferred from one object to another. Concerning the production of sound by a vibrating string, he claims that apart from the fact that *"this vibration is followed by this sound*...we have no idea of it" (E, p. 77). As for palsy and the voluntary control of our muscles by our nerves, he states that they are "to the last degree, mysterious and unintelligible....wholly beyond our comprehension" (E, pp. 66–67). But his profession of ignorance regarding the nutritive powers of bread is the best il-

lustration of the tremendous gap in knowledge between Hume's day
and our own.

> It is confessed that the colour, consistence, and other sensible
> qualities of bread appear not, of themselves, to have any connexion
> with the secret powers of nourishment and support. For otherwise we
> could infer these secret powers from the first appearance of these
> sensible qualities. . .contrary to plain matter of fact. Here, then, *is our
> natural state of ignorance with regard to the powers and influence
> of all objects* (E, p. 37; italics added).

In another passage he says that "neither sense nor reason can
ever inform us of those qualities which fit it for the nourishment
and support of a human body" (E, p. 33).

While it is true that the nourishing properties of bread are not
revealed in its sensible qualities, we are no longer limited merely to
sensible qualities in explaining how or why bread nourishes. Today
there is found on most packages of bread a list of ingredients
including flour, milk, and vegetable shortening, along with nutritional components like vitamins, proteins, and minerals that account, at least partially, for the nutritional value of bread. It is the
responsibility of chemists and pharmacologists in the Food and
Drug Administration to ascertain by chemical analysis and meta-
bolic studies the ingredients and harmful or beneficial properties of
various foods and drugs, thereby *forewarning* the public of their
possible effects. As disappointing as these results may sometimes
be, they do constitute a form of knowledge with *predictable* consequences which Hume claimed "never" could be attained.

The essential reason for Hume's skepticism was his belief that
scientific knowledge was *inevitably* limited to the observable
qualities and conjunctions of objects or, more accurately and paradoxically, to our impressions and ideas of them (for Hume, as for all
the empiricists, no direct knowledge of the world was possible
because our immediate experience of impressions and ideas screens
our perception of the world). This belief, based on his phenomenalistic epistemology, was also dependent upon the limitation,
universally acknowledged in Hume's day, of explanations in terms
of "insensible powers." As he frequently states, "there is no known
connection between the sensible qualities and the secret powers"
(E, p. 33), though again, he is not denying that "secret powers"
exist. In the *Treatise,* for instance, he affirms "that almost in every
part of nature there is contain'd a vast variety of springs and princi-

ples, which are hid, by reason of their minuteness or remoteness" (T, p. 132). Not realizing the possiblity of discovering by experimentation some of these "hidden springs and principles," thereby accounting to some extent for the properties and effects of things, he concluded that they were inevitably "secret," "hidden," and "incomprehensible."

This skepticism regarding possible explanations of phenomena in terms of their underlying causes was not unique to Hume, but prevalent at the time. Anticipating Hume (as well as Kant), Locke, despite his belief in primary qualities, states that

> it seems probable to me, that the simple ideas we receive from sensation and reflection are the boundaries of our thoughts; beyond which, the mind, whatever efforts it would make, is not able to advance one jot; nor can it make any discoveries when it would pry into the nature and hidden causes of those ideas.[9]

Locke also antedated Hume in asserting that "simple ideas... carry with them, in their own nature, no visible necessary connection or inconsistency with any other simple ideas" (Bk. IV, ch. 3, sect. 11).

While Locke denied any "necessary connection or inconsistency" among our simple ideas, Berkeley carried the skeptical argument farther in denying that experience provides any justification for believing in an independent physical reality.

> I say...that it is possible we might be affected with all the ideas we have now, though no bodies existed without, resembling them. Hence it is evident the supposition of external bodies is not necessary for the producing our ideas: since it is granted they are produced sometimes, and might possibly be produced always in the same order we see them in at present, without their concurrence.[10]

If phenomena *could occur* just as they do now *without any* accompanying physical causes, then their occurrence is not evidence for the continuation of these causes, as Hume concluded. In addition, Berkeley anticipates Hume in asserting that "food nourishes, sleep refreshes, and fire warms us...all this we know, not by discovering any necessary connection between our ideas, but only by the observation of the settled laws of Nature" (I, 31). Hume's originality consists in denying any justification for expecting even the continuation of the "settled laws of Nature."

But it was probably the "incomparable Mr. Newton," as Locke described him, who was the principal source and authority of a reserved skepticism, since in a famous passage he appears to limit knowledge to a "deduction from phenomena"—though in many other passages he attests to a deeper understanding of nature in terms of "natural powers" and the "forces of nature."

> Hitherto we have explained the phenomena of the heavens and of our sea by the power of gravity, but have not yet assigned the cause of this power. . . .[because] I have not been able to discover the cause. . .and I frame no hypotheses; for whatever is not deduced from the phenomena is to be called an hypothesis; and hypotheses, whether metaphysical or physical, whether of occult qualities or mechanical, have no place in experimental philosophy.[11] (Brackets added.)

For all his notorious critique of the uniformity of nature and of induction, Hume was actually following out the consequences of a well-established empiricist skepticism. Thus his most eloquent expression of that skepticism reflected the generally acknowledged limits of possible knowledge in the eighteenth century.

> It is confessed, that the utmost effort of human reason is to reduce the principles, productive of natural phenomena, to a greater simplicity, and to resolve the many particular effects into a few general causes, by means of reasoning from analogy, experience, and observation. But as to the causes of these general causes, we should in vain attempt their discovery; nor shall we ever be able to satisfy ourselves, by any particular explication of them. These ultimate springs and principles are totally shut up from human curiosity and enquiry. Elasticity, gravity, cohesion of parts, communication of motion by impulse; these are probably the ultimate causes and principles which we shall ever discover in nature [practically a direct paraphrase of Newton's *Principia*]. . . .The most perfect philosophy of the natural kind only staves off our ignorance a little longer. . . .Thus the observation of human blindness and weakness is the result of all philosophy, and meets us at every turn, in spite of our endeavours to elude or avoid it (E, pp. 30–31; brackets added).

If this expression of the "ultimate" limits of knowledge were the crux of Hume's argument, I doubt many would disagree with him; however, it is *crucial* to remember that his skepticism cuts much deeper, denying not only that we have a knowledge of "ultimate springs and principles," but that *any knowledge we can ever acquire* of the natures or causal powers of objects would constitute a rational justification for believing they would produce the same appearances and effects (under similar conditions) in the future. Yet given the discovery that free hydrogen ions in acids cause litmus paper to turn red, is it just as likely that in the future a liquid manifesting all the properties of sulfuric acid will turn litmus paper blue? Or concerning our knowledge of the atomic–molecular structure of hydrochloric acid (HCl) and sodium hydroxide (NaOH), which when combined form salt (NaCl) and water (H_2O), is it just as possible that in the future they will combine and form potassium sulfate and mercury oxide? Or consider Popper's example of the people in a French village who died from eating their *baguettes*;[12] would we entertain the belief that bread *with the very same ingredients* would have its causal powers miraculously change, thereby producing those tragic effects, or would we expect that some foreign ingredient had entered the bread, isolating ergotism as the cause?

Knowing what we do about the function of DNA in cell replication, is it equally probable that a fertilized human cell will develop into a giraffe? Considering our knowledge of nuclear fission, when uranium atoms are split by neutrons releasing tremendous amounts of nuclear energy, would the world freezing over be just as likely as an atomic explosion? Having constructed a match from chemicals known to ignite under certain conditions, would it be just as plausible that when we strike a match the Tour Eiffel will collapse or that the moon will change its orbit or that the holder of the match will disappear?

Far fetched as these examples are, according to Hume it is just as probable that a familiar object will produce the most bizzare effect as the most expected, because *no connection can ever be established* between the internal natures or causal powers of objects and their effects or properties. Regardless of our knowledge, everything is disconnected! As he unequivocally states: "upon the whole, there appears not, throughout all nature, any one instance of connection which is conceivable by us. All events seem entirely loose and separate" (E, p. 74). Natural connections are not even *conceivable*!

But surely this conception of the world is no longer true. If, in our investigations and explanations of nature *we were* limited solely to the observable qualities of objects, so that we could never experimentally discover, infer, or by modern instruments observe the inherent structures of substances, then Hume's conclusion would be plausible. If we knew nothing more than the visual appearance, acrid smell, and burning sensation of acids, then we could not understand why they turn litmus paper red, a test of acidity. Were our knowledge of hydrochloric acid and sodium hydroxide restricted to sensory observations, we could not explain why they form salt and water when combined. If we had no understanding of atomic or nuclear structures, then how could we produce a controlled nuclear reaction? Knowing nothing of the ingredients of bread or why they have the physiological effects they do, we would be ignorant of why it nourishes. If only the external body could be observed, so that we knew nothing of the difference between striated and unstriated muscle fibers, and even less of neurons and the transmission of nerve impulses to the muscle tissues, then of course how our nerves activate our muscles would be "incomprehensible."

What would be the purpose of building electron microscopes, linear accelerators, and giant optical and radio telescopes with the intent of penetrating more deeply into the inner recesses and outer limits of the universe, if knowledge were limited solely to observable phenomena? Even more decisively, how could we have landed astronauts on the moon, with all the elaborate technology, exact understanding of forces such as gravity and thrust, and precise calculations and predictions of maneuvers required, if our scientific knowledge were as limited as Hume maintained?

If the Humean persists in asking why these discovered structures and causes have the effects they do, we can only reply that an explanation has to be in terms of something (unless we eventually arrive at a bootstrap theory). One can ask for deeper explanations, but even these will have to terminate in the fact that some things or states (such as electromagnetic fields) just have the properties and effects they do. Yet this *does not* concede Hume's thesis! Hume claimed that since we could only observe disconnected sensory qualities, we could never know the "hidden natures" or "secret powers" that produce the observable similarities and sequences that he never denied. If we respond to his skepticism by replying that these observable effects are produced by experimentally discovered and confirmed micro structures, and the Humean persists in demanding, "but why should these microstructures exist with these effects," then we need only reiterate that an explanation will always have to be in terms of something, unless one insists that to explain anything we must know everything, which was not Hume's argument.

Or, we can take the offensive and ask the Humean what would be an acceptable answer? If explaining acidity in terms of hydrogen ions, chemical combination in terms of sharing free electrons, digestion in terms of nutritional value, solidity in terms of strong and weak nuclear forces or the exchange of virtual particles are not explanations, then what would be? If landing men on the moon is not an indication of confirmed inductive generalizations and predictions, what would be accepted as confirmation? A question for which there is no possible answer is not a reasonable question!

Consider the example of a spring–driven, stem watch. A person unfamiliar with the works of such a watch could observe the regular movement of the hands without understanding how or why they rotate: though the watch kept perfect time he would have no reason for expecting it to continue to keep time in the future. But suppose we explain the mechanism of the watch in terms of springs, gears, jewels, stem, and so on, and then remove the case of the watch exposing the works. Would he now have more reason, based on his understanding of the mechanism, for expecting that a wound up watch will continue to run in the future? I think we would agree he would. But suppose the Humean persists and asks, "why should we expect, given the observed mechanism, that it will continue to function in the future as it has in the past?" Supplementing the previous explanation, one can answer in terms of the molecular properties of metals and the laws of dynamics, explaining the tension of springs, the engagement of gears, and the hardness of metals, along with the way in which all these components interact under the usual conditions of gravity, humidity, and temperature to produce the movement of the hands. If the Humean still demands, "why should we expect the atomic–molecular structures and the dynamic laws to persist so that they have the same effects in the future," how are we to reply? What answer would the Humean accept? Is it necessary that we know everything to explain some things? The point is, a reasonable answer to Hume's skepticism has been provided, and any further skepticism is otiose (remembering that Hume's skepticism was dependent upon his belief that we were *inevitably* limited in our explanations of natural occurrences to the *observable* qualities of objects). Knowledge of the world as we know it is possible precisely because phenomena, although existing within certain contexts or conditions, exhibit a degree of autonomy enabling us to explain their occurrence relative to those conditions, without possessing a complete or ultimate knowledge of nature.

Quite simply, then, since Hume's skepticism regarding the uniformity of nature and of induction presupposed an inevitable ignorance of the basic structures and properties of objects which explain how and why they produce the effects they do, this skepticism can be resolved by demonstrating that today we do possess such knowledge, at least to a degree. By examining phenomena under various experimental conditions (e.g., chemical analysis of compounds, isolating infectious virus, and colliding subnuclear particles) to discover their "secret natures" or composition and "hidden powers" or causes, and by constructing theories to interpret the experimental results, we have gained considerable insight into the workings of nature. While undoubtedly we are remote from any final, conclusive, or "ultimate" explanation of most natural phenomena—if such an account is even possible—*this does not mean that the knowledge we do possess has no explanatory significance or predictive value.* Fortunately, we do not have to know everything to explain something. When we can determine the internal composition of subtances and isolate the causes of phenomena to explain, control, and predict natural occurrences, to the extent of synthesizing new materials, preventing or curing diseases, and creating new subnuclear particles, it would be *unreasonable* to doubt our ability to discover *some* of the underlying conditions and physical interdependencies that support our inductive inferences. *Where experimental discoveries and theoretical constructs and frameworks have supplemented our limited and fragmentary sensory observations, our knowledge is more complete and our belief in the deeper interconnected causal matrix producing nature's manifestations more justified.*

Such discovered interconnections attest to a *natural* and a *conceptual* form of necessity, although because they are empirically discovered, they are *a posteriori,* rather than *a priori.* Since both Hume and Kant accepted the traditional assumption that factual knowledge must be *a posteriori* while necessary knowledge must be *a priori* (discovered by reason itself), they excluded the possibility that *natural connections inherent in nature,* although discovered empirically and therefore *a posteriori,* could have some *necessity.* Instead, they attributed the *apparent* empirical necessity to our minds, Hume to a natural propensity of human nature and Kant to a transcendental *a priori* category of the understanding.

Realizing that Hume's critique would undermine the rationality of Newtonian science, since any causal explanations and inductive

generalizations would merely be psychological illusions, Kant introduced the novel idea that the underlying principles of Newtonian
science (along with mathematical truths) were both synthetic and *a
priori*; they consisted of an extension or amplification of knowledge (in contrast to being analytic), but were also necessary and
universal because they were the product of a natural synthesizing or
unifying activity of the mind based on inherent concepts such as
causality. Causal empirical connections do not appear to be necessary because they are reinforced by custom or habit; rather, they are
necessary because our minds are structured to think that way. The
difference between judgments based on a subjective association of
ideas ("judgments of perception") and those constituting objective
scientific knowledge ("judgments of experience"), is that the latter
consist of predicates which, though not implied by the subject
term, are necessarily connected to it by the *a priori* concept of
causality. As Kant states: if I say, "when the sun shines on the stone,
it grows warm," this is merely a subjective judgment based on a
contingent, sequential association.

> But if I say, "The sun warms the stone," I add to the perception a
> concept of the understanding, namely, that of cause, which neces-
> sarily connects with the concept of sunshine that of heat, and the
> synthetical judgment becomes of necessity universally valid. . . .[13]

Believing, as Hume did, that we could have no knowledge of
things as they are in themselves, but only as they appear to us, Kant,
no more than Hume, could foresee that if we knew about the radiant
energy of the sun and the atomic–molecular nature of the stone, we
could explain why sunlight *necessarily* warms the stone by transmitting some of its energy to the kinetic motion of the atoms
composing the stone. Once we realize that the causal powers and
the susceptibility of objects to certain effects depend upon their
atomic–molecular structures, we are justified in attributing a certain *inherent necessity* to nature that underlies our theoretical
interpretations and truth claims. In the words of Harré and Madden:

> The natural necessity in the world is reflected in a conceptual neces-
> sity in discourse about the world. Predicates are bound into ensem-
> bles by virtue of the joint origin of the properties they ascribe to
> things, in the nature of those things. When we think there is a natural
> necessity between manifested properties and hypothesised disposi-
> tions, that is, a real connection via the nature of the thing, then we

are entitled to make a conceptual link, incorporating the power or tendency to manifest the property within the concept of the thing or substance (p. 6).

Furthermore, this interpretation enables us to explain the cognitive status of causality, incorporating the insights of both Hume and Kant, by means of our present knowledge (as described in Chapters I and II); if, as human organisms, we have evolved within a causal matrix of natural events, then it is understandable that our nervous system has been genetically programmed to process sensory stimuli in terms of certain inherent expectations or presuppositions. Thus Hume's notion of human nature having an inherent propensity, when conditioned by repetitive sequences, to expect their recurrence, or Kant's concept of causality as a transcendental category of the understanding that predetermines us to think causally, can be explained as an inherent mode of processing sensory information built into the nervous system due to its evolved response to the causal effects of the world. There seems to be considerable evidence for this in the infant's instinctive expectations and in the prevalence among children and primitive peoples to assume that any occurrence has a cause.[14]

Saul Kripke first rejected the traditional convention that locked analytical necessity to *a pribricity* and factual contingency to *a posterioricity,* when he claimed that *necessary* empirical judgments could be discovered *a posteriori.* According to Kripke, once we have learned that substances have certain intrinsic natures, for example that water is H_2O, that lightning is an electrical discharge, that copper has the atomic weight 63.52 and atomic number 29, and that the diversity of phenomena can be organized into natural kinds based on invariant principles (such as the fact that normally, only members of the same species can procreate), these empirical discoveries can be considered necessary truths, though they are not *a priori.*

> In general, science attempts, by investigating basic structural traits, to find the nature, and thus the essence (in the philosophical sense) of the kind. The case of natural phenomena is similar; such theoretical identifications as 'heat is molecular motion' are *necessary,* though not *a priori.* The type of property identity used in science seems to be associated with *necessity,* not with a priority, or analyticity. . . .[15]

Philosophers who now accept a realistic conception of causality and of scientific theories, and who defend the rationality of theory changes in science, such as Harré and Madden, Putnam, Roy Bashkar, Newton–Smith, Peter Smith, and Michael Devitt, have been influenced somewhat by Kripke's notion of necessary *a posteriori* truths. However, given the contingent nature of scientific discoveries, along with the fallibility of all scientific theories (so that it follows inductively that any current theory, however well-confirmed and established, will eventually be proven false), many philosophers are reluctant to embrace the concepts of "necessary scientific truths," "real natures or essences," and "natural kinds." These latter philosophers, among them the followers of Hume, Dewey, and Wittgenstein, are more apt to defend an instrumentalist interpretation of scientific theories, as well as a more nominalistic or conventionalistic conception of our knowledge of the nature of things.

In the remainder of this chapter, I shall defend the first position, while taking into account the objections of the second. Although all scientific theories are subject to revision and/or falsification, and though undoubtedly there is much more to reality than any of our theories comprehend, so that we must always be cautious in claiming to know the "ultimate," "real," or "final natures" of things, yet the history of theories such as the atomic theory does justify the claim that we do know *certain* necessary scientific truths based on the natures and kinds of things discovered in the world, *relative to this theory.* But since no empirical knowledge (in contrast with the assertions of mystics) can be made independently of any conceptual–linguistic framework, the latter qualification of theory dependence does not undermine the claim. Those who reject the thesis of scientific realism, that the progressive development of scientific theories has depended upon *some* contact and approximate match with at least *some* objective features or dimensions of physical reality, do so, it seems to me, on the grounds that a realist belief would be justified only if we had an absolute, unmediated knowledge of the physical world. But given man's finite, conditional status within the world, such a requirement seems unwarranted.

I do not see how any careful reading of Enrico Cantore's *Atomic Order,*[16] Emilio Segrè's *From X-Rays to Quarks,*[17] Max Jammer's *The Conceptual Development of Quantum Mechanics,*[18] or Horace F Judson's *The Eighth Day of Creation*[19] could fail to convince one that considerable progress has been made in understanding the atomic–molecular, nuclear, and biomolecular structure of matter. Phenomena which were completely incomprehensible, if not unknown, prior to the development of modern atomic and particle physics, are now intelligible. New discoveries have been predicted and confirmed based on experimental evidence and on inferences drawn from theoretical interpretations. More powerful apparatus have been developed to probe nature, explore the implications of theories, and to test hypotheses. The applications of these discoveries have made possible all kinds of new developments, such as nuclear fission and fusion, spectroscopic analysis of astronomical data, molecular biology, plasma physics, laser technology, the use of microchips in computers, and so on. We would have to believe in miraculous coincidences to deny some element of truth in these complementary developments. However, as Cantore states, "Too many philosophers feel inclined toward sweeping generalizations about science without stopping to examine concrete scientific issues in detail" (AO, p. 9).

Although Kripke introduced the notion of *a posteriori* necessary truths in connection with his investigations of modal logic and possible worlds semantics, such truths can only be justified in terms of actual developments in science, such as the atomic theory. The latter, while proposed as early as the fifth century B.C. by the Greek atomists Leucippus and Democritus as a speculative theory to explain the composition, motion, and changes of matter, without violating Parmenides' principle that "nothing can either come to be from nothing nor cease to be,"[20] has in its modern development depended upon the following procedures: (1) inferences based on observed and discovered regularities among phenomena, (2) deductions derived from experimental evidence and theoretical interpretations, and (3) experiments involving chance discoveries as well as deliberate explorations and attempts to test derived predictions. Each of these procedures, directly refuting Hume's claim that "upon the whole, there appears not, throughout all nature, any one instance of connection which is conceivable by us," has been essential to the growth of scientific knowledge.

While it is not possible to present a history of the development of atomic theory in detail, even a brief description illustrates how a web of scientific concepts, inferences, and laws derived from macroscopic and experimental evidence was woven into a theoretical framework with *necessary implications*. The evidence was acquired *a posteriori,* not *a priori* from assumed postulates or self–evident principles, yet the systematic connections of the theory are not merely contingent but exhibit a certain necessity, as indicated by the confirmed conceptual or mathematical deductions from it.

Even observations of phenomena on the macroscopic level suggest the possibility of an atomic structure, as Cantore indicates.

> Clues to the existence of atoms are supplied by everyday experience. Macroscopic substances, though apparently continuous, appear to be divisible into finite, very minute identical particles that contain the chemical properties of the substances themselves. This explains the existence of solutions or homogeneous mixtures (such as sugar in water or gases in an atmosphere). The components intermingle in widely different proportions, while preserving their properties, as it appears from the possibility of recovering the original substances unchanged. Common experience also shows that all specimens of a given chemical substance (for instance, water) have the same properties. This leads to the assumption that all the particles which constitute the substance are homogeneous, indistinguishable and quite independent of their previous history. Chemical substances, however, are most often found in the form of compounds. Thus, in order to understand chemical combinations it would be enough to suppose the elements themselves to consist of indivisible particles, all absolutely identical, with characteristic properties. These particles, remaining unchanged throughout chemical reactions, would then be atoms (AO, p. 19).

Implied in this statement are the very hints and assumptions that led chemists in the eighteenth and nineteenth centuries to the belief that gases, liquids, and material substances are composed of "hypothetical," indivisible, indestructible atoms. Lavoisier, in 1789, clearly stated the two presuppositions essential to the empirical investigations: the conservation of mass and the definition of a chemical element as the ultimate, undecomposable unit in chemical reactions. But it was the fact that the investigators were confronted with certain striking regularities in their analysis of substances, gases, and chemical combinations, proving that "the proportions by weight in which elements are contained in any compound. . .are definite and invariable" (AO, p. 19), and that volumes of gases also react in fixed proportions, that led to the belief that substances and gases were composed of elementary particles possessing a precise mass or weight that combine in definite ratios.

When several hypotheses more accurately describing the nature of the interactions were proposed, nature provided the evidence for choosing among them. Thus Dalton's postulate that elementary gases are always composed of *single* atoms was contradicted by Guy–Lussac's discovery that two volumes of hydrogen gas and one volume of oxygen gas produce two volumes of water vapor, for Dalton's postulate implies that we would end up with twice the number of oxygen atoms we had at the beginning. However, the experimental discovery was consistent with Avogadro's hypothesis that hydrogen and oxygen are composed of diatonic molecules with equal volumes of gases containing equal numbers of molecules: $2H_2 + O_2 \rightarrow 2H_2O$ (cf. AO, pp. 21–23). His assumption, known as Avogadro's principle, that equal volumes of different gases under the same conditions of temperature and pressure contain an equal number of molecules, was proved much later by experiments in crystallography.

> Given a certain type of crystal, we make a careful absolute measurement of the lattice spacing: from it the volume occupied by a molecule can be easily calculated. Then we multiply the molecular volume by the density of the crystal: this gives the mass of a molecule. Finally we divide the gram–molecular weight by the mass of the molecule. The result is the Avogadro number. Needless to say, the value agrees with those obtained by all other independent methods (AO, p. 48).

The exactness of such mathematical calculations, based on precise experimental measurements obtained from independent areas of research, corroborates the belief that empirical connections expressed in physical laws have a certain necessity.

A century before these investigations in crystallography, the hypothetical status of atoms to account for the quantitative regularity of chemical interactions gained indirect observational support from the English botanist, Robert Brown. In 1827, he noted that grains of pollen suspended in a liquid and observed under a microscope display a random motion that could not be attributed to the pollen grains themselves, nor to the effect of anything outside the liquid. He surmised that their irregular motion was caused by the agitation and impact of imperceptible particles within the liquid. Then, in 1905, Einstein published a paper containing a more detailed analysis of Brownlan motion, deriving quantitative predictions for the motion of the pollen grains based on the impact of atoms that were later confirmed in remarkable investigations by the experimentalist Jean Perrin. Again, such complementary confirmations reinforce the belief that the theories describing these atomic structures have some validity.

But prior to the twentieth century, the systematic organization of elements into distinct groups according to their atomic weights and numbers by Mendelyeev, in 1872, lent considerable support to the atomic hypothesis. For the periodicity of properties seemed to be a manifestation of a natural order based on the weights of atoms that hardly could be attributed to coincidence or imagination, especially as Mendelyeev was able to predict the existence of elements, discovered later, to fill in the gaps in the table, and correct previous classifications of elements based on incorrect weights. As he stated, "a true law of nature anticipates facts, foretells magnitudes, gives a hold on nature, and leads to improvements in the methods of research" (AO, p. 39). Quite a different assessment of science from that of Hume!

The millennia–old conception of the atom as the most elementary, indivisible element was then directly challenged by two experimental investigations, one leading to the precise identification of a more elementary particle, the electron, by J. J. Thomson, and the other providing experimental evidence of the internal structure of atoms, by Ernest Rutherford. Both developments illustrate the way experimental inquiries eventuate in the discovery, description, and naming of new particles, thereby shedding light on one of the fundamental questions in the philosophy of language.

To begin with the discovery of electrons, experiments with cathode tubes led to the identification of cathode rays that Michael Faraday described as "visible elementary intermittent charges." Further investigation produced the conviction that the cathode rays were electrons, a name suggested by G. Johnstone Stoney, in 1894.[21] Initially, the name merely designated the glowing rays believed to be charged particles observed in a vacuum tube. Then, in 1895, Jean Perrin, the same physicist who confirmed Einstein's Brownian Motion calculations, proved experimentally that the cathode rays (the term then in use) were negatively charged particles, which became the essential or defining meaning of electrons. Thus, while J. J. Thomson is often credited with "discovering" electrons, their actual discovery preceded him by several years; however, as a result of his experiments presented in a paper in 1897, he was able to confirm the corpuscular nature of electrons, along with the fact that they remained unchanged regardless of the composition of the anode and cathode or of the nature of the gas in the tube. More importantly, he succeeded in determining, by a process of experimental measurements and deduction, the ratio of the electric charge of the electron to the mass, a value of 2.3×10^{17} esu/g (cf. AO, p. 19).

As this example illustrates, certain experimental effects were observed before the name 'electron' was applied to them. Even when the term 'electron' replaced 'cathode rays,' the properties of charge and mass which came to be the electron's defining characteristics were not yet known. Now, however, it is safe to say that 'negatively charged particle' is the standard meaning of 'electron.' Accepting Hilary Putnam's notion of "the division of linguistic labor,"[22] the standard meaning is the sense of the term as understood by the linguistic community at large, though physicists would define an electron more precisely in terms of its charge, mass, and spin. Thus the descriptive meaning of 'electron,' that at first was known only to a few experts, gradually settled into a standard meaning understood by the linguistic community in general that is less precise than the description known to physicists. The meaning of 'electron' has been refined further by Bohr's explanation of atomic specta and by Heisenberg's indeterminacy principle that precludes an electron from possessing a precise position and momentum, and energy and time, at the same moment. But these later refinements do not prevent the word from having a standard meaning in society; if one did not know that an electron was a negatively charged particle he could not use the term properly, leading us to conclude that he did not understand it.

Furthermore, I think this illustrates that Kripke's and Putnam's "causal theory of reference," though it may be supported by considerations in modal logic and other worlds semantics, does not conform to how we actually learn the meanings of words and their correct use. As the example of the discovery of electrons illustrates, no initial experimental effects were "baptized" with the word 'electron,' which identification was then passed on to others through a chain of causal effects. Originally, the word 'electron' was not used at all, being introduced later by someone who was not one of the primary investigators. For a time, 'cathode rays' was used rather than 'electrons,' though gradually the latter replaced the former, and at first 'electron' did not even designate what later became the essential property or standard meaning of electrons, their negative electrical charge. Moreover, there was no specific observable that was being baptized. Thus the naming procedure, far from resembling an initial baptism, can be a somewhat complicated, drawn out, accidental process. For this reason, Putnam's illustration of how scientific terms are learned with his example of "standing next to Benjamin Franklin as he performed his famous experiment. . .[and being told] that 'electricity' is a physical quantity" (p. 200), would seldom be borne out by history. And in my opinion, *what actually takes place ought to take precedence over what philosophers conjecture must be the case to solve certain abstract, logical problems*. The positivists taught us that!

In actual practice, we acquire the name and learn the meaning of technical terms not because of a continuous causal chain, but as a result of reading professional articles and listening to papers or lectures in which the term is used in describing experimental results. There is, of course, a causal aspect to this, as there is in any input of information, yet the essential point is not the *causal effect* of being in proximity to the introducer of the term, but the fact that the reader or listener understands the *meaning* of what is being conveyed, thereby learning the definition and use of the word. In discussions following a lecture, one will often hear the question, "I understand that you call that a so-and-so, but what exactly do you mean by that?"

These considerations also help clarify what seem to me to be inherent confusions and paradoxical consequences of Putnam's notion of meaning in his well-known article, "The Meaning of 'Meaning.'" In attempting to support a realist interpretation of truth, Putnam defends the position that "the nature of things" determines the *ex*tensional meaning of words, not the *in*tensional meaning or psychological state of the individual. In contrast to Wittgenstein and Quine, he is not denying meanings as such: "The reason that the prescientific concept of meaning is in bad shape is not clarified by some general skeptical or nominalistic argument to the effect that meanings don't exist" (p. 216). What he is opposed to is the traditional position that knowing the meaning of a term consists of being in a psychological state, in the sense that remembering or deducing something is being in a psychological state, and that the *in*tensional meaning of a term determines its *ex*tension. C.I. Lewis, for example, claimed that knowing the extension of a term depended upon having a "criterion in mind" that determined the application of the term: "These, then, are the only practical and applicable criteria of common knowledge; that we should share common definitions of the terms we use, and that we should apply these terms identically to what is presented."[23] Putnam denies this on the grounds that it is possible for two speakers to be in exactly the *same* psychological state...even though the extension of the term A in the idiolect of the one is different from the extension of the term A in the idiolect of the other. Extension is *not* determined by psychological state (p. 222).

He holds this view because he believes that a term's extension is *dependent upon the nature of things and thus is independent of what is known by the members of a linguistic community* — at least at any one time. To support this, he discusses a situation in which there is a planet called "twin earth" that has a liquid named 'water' with the same observable or phenomenal qualities as the water on our earth; it is clear, drinkable, an excellent solvent, and capable of being frozen and evaporated. However, in spite of these observable similarities, the atomic–molecular structure of water on the earth is H_2O, while that of the water on twin earth is XYZ, though neither inhabitants know this at the time, 1750. Apart from the consideration whether two substances with an *entirely* different atomic–molecular structure could have *the same* observable qualities and physical properties, Putnam claims that *even before* the people on earth and twin earth knew anything about the molecular structure of what they called water on their respective planets, and thus that what they meant by 'water' (their *psychological states*) was the same, nevertheless, the *extension* of 'water' was different because the water on each planet was in fact different. As he says,

the extension of the term 'water' was just as much H_2O on Earth in 1750 as in 1950; and the extension of the term 'water' was just as much XYZ on Twin Earth in 1750 as in 1950. [People on the two planets]...understood the term 'water' differently in 1750 *although they were in the same psychological state,* and although, given the state of science at the time, it would have taken their scientific communities about fifty years to discover that they understood the term 'water' differently (p. 224; brackets added).

Making allowance for his unusual use of "understood" (in that it is difficult to see how something can be "understood" before it is known), it is still paradoxical to claim that in 1750 the term 'water' had a different *ex*tensional meaning *for* the people on earth and twin earth because of the difference in atomic structure, *although they knew nothing of the difference at the time.*

Wouldn't it be more reasonable to describe the situation in 1750 as one in which the word 'water' had the same meaning, *ex*tensional as well as *in*tensional, for the people of both planets because it was not yet known that what each designated as water had a different atomic structure? *After* the discovery of the different atomic structure of the 'water' on twin earth, the people on both planets might still retain the word 'water,' but owing to their additional knowledge, the term would acquire different intensional as well as extensional meanings because of the unlike natures of the two bodies of water. The word 'water' would still connote the same observable properties but now include, in addition, the sense of the atomic–molecular structure. In ordinary discourse, the sensible properties would predominate, but if it were a question of deciding whether or not samples of water on the two planets were identical, the atomic structure would be the crucial characteristic. We have a similar situation today as regards genuine and artificial diamonds. People buy the artificial ones for their resemblance to real diamonds and their lower prices, but their actual value is based on their internal structures, as well as on certain physical properties dependent upon those structures.

I think Putnam's position can be clarified by keeping in mind the following distinctions: (1) the world of objects as they appear to human beings, (2) the inherent natures of objects as they exist independently of us, (3) the meanings words have for the experts, (4) the standard meanings of terms in society at large, and (5) the meanings words have for each individual. Putnam is correct in asserting that the intensional meanings of words for each individual, and even the standard meaning accepted by society, may not correspond to the exact extension of the term, which is determined by the nature of things as discovered by experts. As he says,

> the extension of a term is not fixed by a concept that the individual speaker has in his head...both because extension is, in general, determined *socially*...and because...[t]he extension of our terms depends upon the actual nature of the particular things that serve as paradigms, and this actual nature is not, in general, fully known to the speaker (p. 245).

Unfortunately, Putnam draws two mistaken conclusions from this: (1) that *in*tensional meaning does not even establish the *ex*tension of a term for *individual* users of the term, and (2) that the extension of terms is (or can be) independent of what *anyone knows* at a particular time. Regarding the first conclusion, even though a *particular* individual's understanding of a term (unless he

happens to be the world's expert) does not establish the standard
extensional meaning of the term, it does *not* follow that the extension of the term *for that person* is not determined by his or her
conception of the meaning. Though the meaning of a word for an
individual may not coincide with the standard meaning, still, its
meaning *for him* determines the reference for *him;* only *he* can
correctly or incorrectly identify the referent on the basis of *his*
conception. We can always use terms in "a loose sense," but sooner
or later the exactness of our conception will be tested by our ability
to identify the referent. As Putnam says, suppose a "speaker points
to a snowball and asks, 'is that a tiger?' Clearly there isn't much
point in talking tigers with *him*" (p. 248).

As for the second erroneous conclusion, that the extensional meanings of terms are (or can be) independent of what is known by the
community (e.g., the people on earth and twin earth in 1750), this
leads to the following dilemma, each horn of which is problematic.
Either we have to decide somewhat arbitrarily at some point in time
that *someone* possesses a knowledge of the *real* natures of things
that fixes the extensional meaning of the term, or we must maintain
that the extensional meaning is inherent, invariant, timeless, but
unknowable because it refers to the actual natures of things regardless of what is known by man. Putnam asserts that the extensional
meaning of 'water' in 1750 was different from what was believed by
the people on earth and twin earth because *later* it was discovered
that the respective waters have different internal structures. But,
unless he assumes that the later discovery is final and conclusive, he
would have to acknowledge that further research might disclose an
even more basic nature. In other words, our claim that the extension of 'water' in 1750 must be different from what was known then
because of what we know now, could apply to us today if later
someone discovers that the structure of water is other than H_2O, *ad
infinitum*. What would justify us in accepting a particular conception at any one time as true? To avoid this conclusion, we could
maintain that since extension is determined by the *ultimate* natures
of objects, and since we can never be certain that we know these
ultimate natures, extension transcends what can be known, referring to unknowable things in themselves. But this would imply that
Putnam is a Kantian.

Surely in our present understanding of terms such as 'water' we
do not intend them to refer to the natures of things as they will be
disclosed in some ultimate Peircean future, but as their natures are

known today, to the best of our knowledge. Even when using the term 'quarks,' scientists conceive of these subnuclear particles as possessing such conjectured properties as up–down, bottom–top, charm, and color, not as they will be known (if at all) years from now. What justifies a realist interpretation of science is the fact that our theories do explain phenomena, that they do lead to predictable confirmations, that these confirmations occur in different areas of research and thus do reinforce each other, and that the natures of things as experimentally discovered do force us to revise our theories to make them conform to the evidence—not the assumption that our scientific terms refer to the absolute natures of things as they exist in themselves. We believe that our current network of scientific theories captures some of the inherent properties of objects, and therefore has some approximate truth, otherwise we could not account for its explanatory power; yet we never know to what extent other properties elude us which, if known, would result in quite a different web of beliefs. As Putnam says:

> It is beyond question that scientists use terms as if the associated criteria were not *necessary and sufficient conditions,* but rather *approximately* correct characterizations of some world of theory-independent entities, and that they talk as if later theories in a mature science were, in general, *better* descriptions of the *same* entities that earlier theories referred to. In my opinion the hypothesis that this is *right* is the only hypothesis that can account for the communicability of scientific results, the...[acceptability of] scientific theories...and many other features of the scientific method (p. 237; brackets added).

I fully agree with this excellent statement, but disagree that it implies the theory of meaning advocated by Putnam. Even his notion of "stereotype" or of "paradigm" consists of using other words for what is usually meant by a cluster of properties or intensional meaning.

I broke off the description of how the initial evidence for the atomic theory exemplifies the manner in which empirical regularities induce experiments that, in turn, give rise to theories with *necessary* implications or connections to examine how the discovery of electrons illustrates the way new scientific terms enter into language and acquire their meanings. To return to the second experimental inquiry, Rutherford's, that led to the rejection of the ancient conception of the atom as an ultimate, indivisible element of nature, no recapitulation of the reasoning process based on the experimental evidence could fail to illustrate *the interdependence of empirical evidence* and *the interconnectedness of the theoretical inferences and deductions.*

Prior to Rutherford's exploratory atomic experiments, various physical investigations had convinced scientists that material bodies are composed of large numbers of electrons, but since electrons are negatively charged, with hardly any mass, while material bodies are usually neutral in charge, with a large mass, they deduced that matter must consist of other particles possessing a large mass and positive charge. Because the empirical evidence did not indicate the arrangement of these elements, a number of theories were advanced describing possible atomic structures. Thus it was necessary, somehow, to probe material bodies to derive evidence indicating which theory was correct. The method that seemed to have the greatest likelihood of success consisted of attempting to pry into an object's composition with some kind of particle, with the hope of eliciting evidence of its structure. This required that the probe possess a positive charge with sufficient energy to penetrate the atom and interact with the postulated positively charged, massive particles without being deflected by the electrons.

The experimental stage was set for Rutherford who, in 1898, had concluded that two kinds of radiation were emitted from uranium which he designated alpha and beta. It was established a few years later that the beta radiation consisted of cathode rays (the electrons mentioned previously), while further experiments by Rutherford confirmed that the alpha rays were positively charged particles, in fact, ionized helium. Luckily, these alpha particles possessed exactly the properties needed to probe the atom: positive charge, large mass, high energy, and extremely high velocity. Thus the required information regarding the interior of atoms might be obtained by directing the alpha radiation at thin metal foils so that the scattered particles could be observed on a florescent screen, and their angular deflection measured.

When Rutherford performed the experiment using gold foil, he expected the angle of deflection to be slight, approaching zero; instead, he was astonished to find that while almost all the particles passed through the foil unimpeded, a few were repelled at large angles, some even deflected backwards! Reflecting on these unexpected results, Rutherford soon realized their implication regarding the structure of the atom; namely, that it consisted of a massive, positively charged, point–like center which he called the "nucleus," around which circled electrons, maintained in their orbits by the counteracting effects of the electromagnetic attraction between them and the positively charged nucleus, and the centrifugal force due to their angular momentum.

Having arrived at the planetary conception of the atom, Rutherford then derived his famous formula from the dimensions of the thickness of the foil, the initial energy and the angles of deflection of the alpha particles, and the inverse square law of electrostatic forces. He deduced the nuclear size and charge from measurements of the interactions of the alpha particle with the nucleus. Finally, he offered "a physical interpretation of the atomic number in terms of the nuclear charge," when he found that the values of the electronic charge coincided with the atomic number of the particular material composing the foil, surmising "that the atomic number was the number of elementary charges carried by the nucleus" (AO, p. 65). These results further illustrate the importance of converging evidence.

Continuing his experiments with alpha particles, though this time using nitrogen gas as his target, Rutherford detected the first nuclear disintegration, confirming the alchemist's dream of the transmutation of elements. He coined the term 'proton' (first one) to designate the nucleus of an atom of hydrogen gas, inferring that the number of protons in the nucleus (which in a neutral atom equals the number of electrons in the outer orbits) is characteristic of a chemical element, determining its place in the Periodic Table. In this way, he was able to account for Mendelyeev's periodic organization of the elements, although a more complete explanation would be provided by Bohr's more advanced Periodic Table.

Having successfully identified the protons in the nucleus, in 1920 Rutherford predicted the existence of a second nuclear particle which he called 'neutron' to account for the fact that most atoms have a mass or weight twice that of the combined masses of their protons and electrons. The existence of this new particle, neutral in charge (thus preserving the electrical neutrality of the atom) with a mass comparable to the proton, was confirmed twelve years later by James Chadwick, proving another illustration of successful predicttions, based on indirect evidence, of the inherent structures or natures of material bodies. The baptism theory of Kripke and Putnam is again belied, since Rutherford's introduction of the term 'neutron' was based on *theoretical inferences* from the experimental data, as so often occurs in science, rather than on pointing to a specific particle. Given what are recognized as *necessary* physical requirements, concepts like the neutron are introduced and defined to provide a mechanism or rationale in agreement with these specifications (other examples are the 'mesons,' 'gluons,' 'quarks').

During the time that Rutherford was conducting his remarkable ground-breaking experiments, other extraordinary developments were occurring that shook the foundations of physics, such as Planck's introduction of a universal quantum of action to account for the anomalous blackbody radiation, and Einstein's conception of light quanta or photons to explain the photoelectric effect. Meanwhile, Bohr, who had gone to Manchester to study with Rutherford, became intrigued by certain problems pertaining to the latter's "Saturnian model" of the atom, such as its stability and radius, and the fact that all atoms of a substance seemed to be the same. While Planck's quantum of action was believed to be a clue to the solution to these problems, it was Balmer's formula, depicting the frequencies of the spectral lines of hydrogen, that led Bohr to the insight that the hydrogen spectra could be explained in terms of electron jumps between various privileged orbits or stationary states.

Utilizing Planck's constant and Einstein's correlation of quanta of energy with radiational frequency, Bohr derived a formula that related the frequency or energy of the spectral emission to the energy lost by an electron dropping to a lower stationary state. Although stationary orbits and discontinuous energy transitions were inconsistent with classical mechanics, by adopting these conceptions Bohr was able to solve the atomic problems as well as derive exact values for the emission spectra of the hydrogen atom. This conception of stationary electron shells, incorporating research in radioactivity, spectroscopy, quantum theory, and atomic physics, provided a precise description of the electron orbits which, in turn, helped to explain the atom's chemical properties. Einstein describes in vivid terms the effect these singular achievements had on him:

> That this insecure and contradictory foundation was sufficient to enable a man of Bohr's unique instinct and tact to discover the major laws of the spectral lines and of the electron–shells of the atoms together with their significance for chemistry appeared to me like a miracle—and appears to me as a miracle even today. This is the highest form of musicality in the sphere of thought.[24]

Subsequent developments in atomic research consisted, first, in deriving a formalism for quantum mechanics, achieved by Schrödinger's wave equation and Heisenberg's matrix mechanics, then in combining quantum mechanics with relativity theory, accomplished by Dirac in his field theory of quantum electrodynamics, followed since mid-century by intensive experiments in particle physics. By colliding highly accelerated, high energy particles such as protons and electrons with subatomic target particles, phyicists have been able to decompose the latter and create new particles from the components of the collision plus the energy of velocity. A new table has been devised organizing the newly discovered or created particles under the categories of Baryons, Mesons, and Leptons based on such exotic properties as spin, mode of decay, and average lifetime, along with the more traditional magnitudes of mass and charge. Presently, there is an elaborate theory of even more basic particles, the quarks, to explain the properties, interactions, and nuclear forces of the hadrons and leptons, and a massive hunt is underway to prove the existence of quarks.[25]

The culmination of these investigations, the current atomic theory, has been called the most powerful theory ever devised by man. Hardly an area of physical research has not been enriched by it. Given such an achievement, it is time to discard Hume's outmoded conception of scientific inquiry as limited to the observable qualities of objects, and scientific knowledge as consisting of completely disconnected, subjectively associated conceptual or semantic elements that have absolutely no factual significance or referential truth, and therefore are entirely devoid of inductive or predictive value. Not only have scientists been successful in discovering the subatomic structure of matter (Hume's "inner natures") and in developing a powerful theory of explanation and prediction, greater confidence in the representational validity of the atomic-molecular theory has been considerably enhanced recently by computer simulations of three-dimensional molecular structures. I previously mentioned the beautiful image of the DNA molecule produced by computer graphics, and now Herbert A. Hauptman and Jerome Karle have been awarded the Nobel Prize in Chemistry for their "direct method" (a mathematical technique for interpreting X-ray crystallographic patterns) of determining the three-dimensional arrangement of atoms within crystals. As a result of such progress, the burden of proof has shifted to those who deny scientific realism. The human condition is no longer the Humean condition!

Furthermore, even this brief description of the origins of the atomic framework does not support a completely holistic interpretation of scientific theories, one that claims that the meaning of scientific terms or concepts is so determined by their role within the theory that any radical changes in the theory irretrievably alters their meaning. Being thus dependent upon the context of the theory for their meanings, the words recurring in transformed theories acquire new meanings and, therefore, refer to different entities; consequently, the new theory is incommensurable with the old. As Kuhn states:

> To make the transition to Einstein's universe, the whole conceptual web whose strands are space, time, matter, force, and so on, had to be shifted and laid down again on nature whole....Communication across the revolutionary divide is inevitably partial.[26]

Or, as he says more emphatically, "the proponents of competing paradigms practice their trades in different worlds" (p. 150).

It is, however, exaggerated to claim that theories by themselves impose meaning on their basic semantic units, as if theories predate their components. Rather, as we have seen in the introduction of such basic scientific terms as 'electron,' 'proton,' and 'neutron,' theories are gradually constructed from more fundamental concepts whose meanings (based on the prevalent linguistic background) are not determined by their role in a theory still in the process of formation, but are dependent on experimental evidence. Introduced to explain certain experimental results, these terms are not explicable by "meaning postulates" but by the experimental data, and thus acquire a *somewhat* specific and independent extensional reference and intensional meaning. As theories develop, the network of implications and the contextual dependency of meanings increase, so that evidence becomes more theory–dependent and theories more impervious to revision; yet the intransigent nature of empirical data or experimental results exacts a certain accommodation: for instance, Kepler's elliptical orbits, Planck's quantum of action, Einstein's invariant velocity of light, Bohr's stationary electron orbits and quantum jumps, and the double helix structure of DNA. However tightly woven a theory, new experimental evidence can rend the fabric, requiring some reweaving of the strands. This interaction and accommodation between established theories and unexpected experimental results grounds the meanings and the references of our theories in the world, however approximately.

New investigations lead to refinements in the meanings of concepts either in terms of more exact properties and magnitudes, or by discovering, at a deeper level, the composition, structure, and causes of more familiar phenomena. Thus the concepts of discourses vary with the level of investigation and explanation. Despite the misgivings of some philosophers, from my observation scientists can pass from one theoretical framework or level of inquiry to another without any special problem, whether it be from the framework of the macroscopic world to that of Newtonian mechanics, or from the latter to relativity theory. Einstein, for example, never characterized classical and relativity mechanics as being in "different worlds," but saw the difference as due to the limited or approximate application of certain concepts. This also was the view of Heisenberg: "we can never know beforehand which limitations will be put on the applicability of certain concepts by the extension of our knowledge into the remote parts of nature. . . ."[27]

Although the concepts of space, time, and mass, as they are defined in Newtonian science, do appear to be absolute, invariant, intrinsic metrics within the context of our usual experience with its limited distances and velocities, once we recognize that light has an invariant velocity, it follows that these metrics must vary with the coordinate system from which they are measured. If light is constant among observers in relative motion, then one can account for this invariance by attributing a contraction of measuring rods and a slowing down of clocks (with an increase in mass) to the coordinate system moving relative to another. These effects will not be manifested under the usual conditions, but they will be pronounced as the velocity of the system approaches that of light (or it enters into stronger gravitational fields). Einstein described Newtonian mechanics as applying within the "limiting conditions" wherein light tends to be instantaneous relative to solar distances and velocities. Under these conditions, the Lorentz transformation equations reduce to the Galilean transformation equations.

Thus Einstein did not view his theory and Newtonian mechanics as incommensurate and the meanings of space, time, and mass as entirely different in each system. In relativity mechanics, space-time still designates the time it takes a light signal to traverse a distance, understood in the traditional sense, though the interval will be measured differently by observers in relative motion; and though mass increases with velocity, it retains the Newtonian properties of resistance to motion and the source of the gravitational field or of the curvature of space–time. Like the electron, which no longer is a particle in the classical sense, since it now has a wave aspect and an indeterminate spatial and temporal location, but still retains the essential property of negative electric charge, the classical concepts of space, time, and mass retain a certain continuity of meaning in relativity theory despite their revisions. Just as we refer to individuals by name and then think of them in terms of various stages of development, traits, and roles, so scientists refer to certain entities by their traditional names and then think of them in various ways, depending upon the experimental or theoretical context.

Even under the most radical or revolutionary theory changes, is it likely that *all* the descriptive or defining properties of *all* of the terms change, along with their syntactical, logical, and semantic connections, so that both the content of the theory, as well as its ontology, become entirely different? Were this the case, the new theory would be *completely unintelligible* because nothing in it would make sense in terms of what was previously known. But the intelligibility of theory change can be preserved if we acknowledge the invariance or partial invariance of some of these aspects; that is, that some of the words, or some of their cluster of descriptions, or some of the semantic connections remain unchanged.

In conclusion, since proper names, natural kind terms, physical magnitudes, and theoretical concepts have acquired their meaning within a linguistic community, that meaning is determined by the knowledge and beliefs of at least some members of the community. Although not all the members understand the same things by the same names, terms, and concepts, there must be some similarity of meaning for unambiguous communication to be possible. The meaning associated with a term consists of a suitable number of descriptions believed by the community to be true of the referent, but the descriptions need not be analytical or invariant, since each is defeasible and therefore can change with time. These standard descriptions, along with the appropriate syntactic markers, specify the referent of the term within the linguistic community. Any entity possessing a suitable majority of the descriptions is necessarily the referent of the term. As most descriptions are based on observation or experimental inquiry, they are a function of the causal effects of the referent as interpreted by experts. Moreover, because the descriptions are intended to be as specific as possible, they should designate the essential features or real natures of the referents, relative to the knowledge available at the time. Thus while all intensional meaning is relative to some conceptual–linguistic framework, and therefore is theory–dependent, in so far as extensional meaning depends upon the real characteristics and natures of things, knowledge is also realistic. Theory change is usually progressive and rational rather than incommensurable because there is continuity among some of the appropriate descriptions of a term as used by successive inquirers. Verisimilitude is dependent upon our beliefs progressively reflecting more of the realistic features of the world, a fact that is historically evident, though it is unlikely that any of these beliefs will turn out to be ultimately or unconditionally true within a broader or deeper context.

Notes to Chapter IV

1. With permission of the Editor, this chapter is based on an article, "A Reasonable Reply to Hume's Skepticism," *The British Journal for the Philosophy of Science,* December, 1984, pp. 359-374.
2. Cf. Richard H. Schlagel, *From Myth to the Modern Mind: The Origins and Growth of Scientific Thought,* Vol. I, *Animism to Archimedes* (Bern: Peter Lang, 1985), ch. XIII and the Epilogue.
3. P. F. Strawson, *Philosophical Studies,* Vol. 9, No. 1-2, p. 20ff. Quoted from Karl Popper, *Objective Knowledge* (Oxford: Clarendon Press, 1972), p. 11.
4. W. V. Quine, "Epistemology Naturalized," in *Ontological Relativity and Other Essays* (New York: Columbia University Press, 1969), p. 72. Brackets added.
5. L. A. Selby-Bigge (ed.), *Hume's Enquiries,* 2nd ed. (Oxford: Clarendon Press, 1902), p. 38. Subsequent references to this work will be indicated in the text by E, followed by the page number.
6. R. Harré and E. H. Madden, *Causal Powers* (Oxford: Basil Blackwell, 1975), p. 50. Subsequent references to this work will be indicated in the text by page number.
7. E. S. Haldane and G. R. Ross (eds.), *The Philosophical Works of Descartes,* Vol. I (Cambridge: Cambridge University Press, 1967), p. 162.
8. L. A. Selby-Bigge (ed.), *Hume's Treatise* (Oxford: Clarendon Press, 1888), pp. 161-162. Subsequent references to this work will be indicated in the text by T, followed by the page number.
9. John Locke, *An Essay Concerning Human Understanding,* Bk. II, ch. 23, sect. 29.
10. George Berkeley, *The Principles of Human Knowledge,* 2nd ed., Part I, sect. 18.
11. Isaak Newton, *Principia,* Vol. II, trans. by A. Motte (Berkeley: University of California Press, 1962), pp. 546-547.
12. Karl Popper *Objective Knowledge* (Oxford: Clarendon Press, 1972), p. 11.
13. Immanuel Kant, *Prolegomena to Any Future Metaphysics,* revised Carus trans. (Indianapolis: Bobbs-Merrill, 1950), p. 49, f.n. 3.
14. Cf. W. V. Quine, "The Nature of Natural Knowledge," in Samuel Guttenplan (ed.), *Mind and Language* (Oxford: Clarendon Press, 1975), p. 69.

15. Saul Kripke, *Naming and Necessity* (Cambridge: Harvard University Press, 1972), p. 138.
16. Enrico Cantore, *Atomic Order: An Introduction to the Philosophy of Microphysics* (Cambridge: M.I.T. Press, 1969). Subsequent references to this work will be indicated in the text by AO, followed by the page number.
17. Emilio Segré, *From X-Rays To Quarks* (San Francisco: W. H. Freeman, 1980).
18. Max Jammer, *The Conceptual Development of Quantum Mechanics* (New York: McGraw–Hill, 1966).
19. H. F Judson, *The Eighth Day of Creation* (New York: Simon & Schuster, 1979).
20. Cf. Richard H. Schlagel, *Animism to Archimedes, op. cit.,* ch. X.
21. Emilio Segrè, *op. cit.,* p. 13.
22. Hilary Putnam, *Mind, Language and Reality,* Vol. 2 (Cambridge: Cambridge University Press, 1975), p. 228. Subsequent references to this work will be indicated in the text by the page number.
23. C. I. Lewis, *Mind And The World Order* (New York: Dover Publications, 1929), p. 76.
24. Albert Einstein, "Autobiographical Notes," in *Albert Einstein: Philosopher-Scientist,* ed. by Paul A. Schilpp (Evanston: Library of Living Philosophers, 1949), Vol. VII, pp. 45–47.
25. Cf. Harold Fritzsch, *Quarks* (New York: Basic Books, 1983).
26. T. S. Kuhn, *The Structure of Scientific Revolutions,* 2nd ed. enlarged (Chicago: University of Chicago Press, 1970), p. 149.
27. Werner Heisenberg, *Physics and Philosophy* (New York: Harper & Row, 1958), p. 85.

CHAPTER V

THE WEB OF HUMAN KNOWLEDGE[1]

BEFORE leaving this expository account of the acquisition of knowledge and turning to normative questions pertaining to the nature and criteria of truth, I shall attempt in this chapter to extract and synthesize the consequences of the previous analyses, in the realization that it is one thing to criticize the views of others and another to offer better alternatives or more adequate solutions to the same problems. When charged with being more critical than constructive, Morris Cohen responded that it was not the task of Hercules to refill the Augean stables once he had emptied them, but this cynical reply overlooks philosophy's equally important constructive function. It is essential that the negative effects of an unduly analytical conception of philosophy be balanced by the positive effort to propose constructive solutions to the persistent or recurrent problems of philosophy, in the tradition of Locke, Kant, Dewey, Russell, Lewis, and Quine.

As previously documented, current neurological evidence opposes the traditional conception of man as an ontologically distinct spiritual or mental substance harbored in the body, indicating instead that a human being is an exceedingly complex neurophysiological organism whose initial responses to the world have largely been preestablished by evolved genetic endowments. As early as Aristotle, it was realized that without the inherent capacity to discriminate among sensory qualities and perceptual forms knowledge would be impossible. Today, empirical evidence indicates that, in addition, infants possess an innate capacity for depth perception and are preprogrammed to respond more positively to some visual patterns than to others. One can conjecture, too, that cognitive abilities that enhance survival, such as a disposition to assent to inductive generalizations[2] and a proclivity towards truth,[3] are genetically endowed—however modifiable by subsequent experience. Furthermore, our present understanding of the elaborate neuronal functions of the central nervous system indicates they do not simply transmit or duplicate peripheral stimuli, but encode, process, and integrate the sensory information into structured percepts, even prior to the sensory–motor activity of learning.

At this level of experience, primates display more or less the same cognitive capacities as humans. Though there is still considerable controversy as to what constitutes a linguistic system and language competence, only the most recalcitrant Cartesian could deny, given the current evidence, that chimpanzees are capable of signing in the American Sign Language and, therefore, of communicating. In addition, any lingering conviction that the species differences between apes and humans are so great as to preclude any similarities in behavior has been convincingly refuted by the field observations of Jane Goodall. Nevertheless, granting these genetically endowed resemblances of behavior pointing to a common ancestry, only man has acquired the capacity to speak and write, abilities that have made possible his unique intellectual development and rich cultural attainments. For unlike animals whose behavior consists primarily of reflex reactions and conditioned responses as elicited by concurrent sensory activation—sounds, smells, and moving images—man's behavior is almost entirely mediated by signs, symbols, and meanings which enable him to transcend the limits of direct stimuli. Everything known as human culture constitutes this mediation: religious symbols, ritual, and beliefs; literature, poetry, and drama; music and mathematics; science and technology; along with ordinary language and common sense knowledge.

Just as our behavioral responses depend upon neurophysiological structures and functions (respiratory, digestive, motor, sensory, etc.) a growing body of evidence indicates that man's linguistic abilities also depend upon inborn neural endowments. The proclivity of infants to learn a language; the flexibility, adroitness, and appropriateness of their understanding and use of novel expressions; the universal sequential stages of linguistic development; the preference for certain word orders and avoidance of expected errors; and the evidence uncovered recently of a universal syntactic structure underlying all creole languages strongly support the hypothesis of a species–specific, innate language–acquisition device as advanced by Chomsky.

If this be the case, then our understanding of the learning and nature of language need not be confined to Quine's naturalistic–behavioristic interpretation, which tends to denigrate meanings, claiming that "[l]anguage is a social art which we all acquire on the evidence *solely* of other people's overt behavior under *publicly recognizable circumstances.*"[4] While language learning *is* cued to observed entities and occurrences as reinforced by adults, a growing body of evidence suggests that the infant also has an inherent ability to categorize sounds, along with an innate schematism possessed by the language–acquisition device that generates rules for correlating surface structures with deep structures, and pairing phonetic with semantic units. In a recent *Scientific American* article, Peter Eimas states that current research, rather than supporting the view that the child's rapid acquisition of linguistic skills can be explained in terms of "such simple psychological principles as conditioning and generalization" (principles relied on by Quine), the experiments

> carried out by my colleagues and me at Brown University and by other investigators elsewhere have supported a different explanation, one derived from the view, of which the linguist Noam Chomsky is the most notable exponent, that inborn knowledge and capacities underlie the use of language. In studies of speech perception by infants we have found these young subjects are richly endowed with innate perceptual mechanisms, well adapted to the characteristics of human language, that prepare them for the linguistic world they will encounter.[5]

Positing such inherent endowments as the capacity to "detect discrete phonemic categories" described by Eimas, and Chomsky's linguistic mechanism for generating grammatical rules would account for the infant's ease and rapidity in learning a langugae, as well as the manner in which the learning occurs. For in contrast to

Quine's thesis that language learning initially depends entirely upon reinforcement and conditioning, parents that I have talked with who have observed their children's acquisition of language closely, have remarked that they are often surprised by the infant's use, in a new or unusual way, of an expression it has heard previously. Having merely heard the word used by adults, the infant will utilize it in a context that has not been subject to deliberate reinforcement and conditioning. This ability of infants to "pick up" word uses and meanings on the basis of just hearing sounds around them may be as important an aspect of later language learning as the intentional attempts of parents to teach or condition infants to use words in a certain way. Such an ability, noticeable among adults, would seem to depend on some inherent capacity to assimilate correct word uses and meanings, apart from external reinforcement. This would illustrate one of the characteristics of human language usage emphasized by Chomsky, namely, freedom from stimulus control.

In addition, these innate abilities would explain certain current difficulties exhibited by the child due to the necessity of modifying inborn phonetic discriminations and possible grammars to fit the specific phonemic units and grammar of its particular natural language. Finally, if language acquisition is assumed to be based on innate endowments as well as external conditioning, this would account for some invariant features of natural languages presupposed by their translatability, despite their obvious differences. For regardless of the phonetic and syntactical variations among Indo–European, Oriental, and American Indian languages such as Hopi, they are intertranslatable to a considerable degree.

The issue of translatability raises the question whether "observation sentences" actually play the crucial role assigned to them by Quine in his epistemology and theory of language learning. For though he accepts the failure of Carnap's program of reconstructing our knowledge of the external world from sense data and reducing scientific statements to assertions about observations, he still maintains the "unassailability" of "[t]wo cardinal tenets of empiricism," as we saw in Chapter III: "that whatever evidence there *is* for science *is* sensory evidence...[and] that all inculcation of meanings of words must rest ultimately on sensory evidence."[6] But as I argued previously, neither statement is "unassailable" today. Due to the influence of such philosophers as Wittgenstein, Hanson, Sellers, Popper, Feyerabend, and more recently, Churchland, it is now generally conceded that *all* empirical assertions are dependent for their meaning and use on some conceptual-linguistic framework, and hence are theoretical and fallible. In contrast, Quine offers several examples of "observation sentences," such as 'this is red' and 'it is raining,' which he claims are *not* subject to the indeterminacy of translation. However, both these expressions can be rendered in other ways so as to convey a different conception of the state of affairs: for example, 'this is red' could be expressed as 'this reds,' 'this shows red,' 'this fires,' or 'this is blood-like,' while 'it is raining' could be stated as 'the sky waters,' 'the clouds wet,' 'the heavens weep,' or 'wetting occurs.' Contrary to Quine's conviction, these different formulations would make cross-cultural, interlinguistic translations and assent of even observation statements questionable, granting an 'identity in the range of sensory stimulation.'[7] Thus if observation statements can be formulated differently, they too are subject to indeterminacy of translation. To modify Neurath's epigram, we are adrift in the sea of language with no firm bedrock to moor on.

Accordingly, the empiricist-positivist thesis of an ultimate cleavage between sensory data and theory, carried over in Quine's graphic assertion that an "observation statement has an empirical content all its own and wears it on its sleeve," in that the meaning of such statements coincides with their direct sensory evocation, is now generally rejected. As Churchland states:

> To acquire a specific language is, among other things, to learn to process peripheral information into the categories that language provides. The causal processes initiated at the periphery come to terminate in the utterance of specific sentences within that language....To acquire a specific language then is indeed to come to share in a specific view of reality. The informational matrix the language embodies comes to shape the processing of peripheral information "from the top down", as it were....[8]

We should also realize that the correlation of observation statements with sense stimuli (in contrast to perceptual objects) is itself not a given fact, but the result of a theoretical interpretation and explanation (which Aristotle would have considered odd). The belief that we are directly acquainted with what impacts on our senses, and that these sensory "impressions" are the prior, indubitable foundation of knowledge, is the legacy of the British Empiricists. It is this legacy of sensory foundationalism that has recently been challenged and rejected by many philosophers.

While it is true that some statements are more directly cued to sensory observations than others, this does not mean they are indubitable or that they ultimately command any greater epistemic certainty. As Feyerabend asserts, "experience arises *together with* theoretical assumptions, *not* before them,"[9] hence *all* experience is theory-laden and susceptible to reinterpretation. As far as science is concerned, where the evidence is more obviously dependent upon the theoretical interpretation including the mathematical formalism, Quine's assertion "that whatever evidence there *is* for science *is* sensory evidence" is highly problematic. In the Michelson-Morely experiments, for example, was the absence of interference patterns in their use of the interferometer sensory evidence? Or to cite the case described by C. I. Lewis,

> when the eclipse-photographs which figured in the discussion of relativity were examined, the question was raised whether the star-displacements as measured on them represented simply the bending of light-rays or were due in part to halation of the sensitive film. Thus, for a moment at least, so fundamental a problem as that of abandoning the categories of independent space and time was intertwined with the question whether the position of dots on a photographic plate represented authentic star-photographs or was due to something which took place inside the camera.[10]

As Lewis implies, the position of the dots on the photographic plate alone hardly settled the issue. Similarly, without the theory of the "eightfold way," the interaction and disintegration of subatomic particles, as indicated by the delicate traceries left in a liquid hydrogen bubble chamber photograph, would appear as nothing more than intricate doodles. We realize today that what is meant by "sensory evidence" in science, as Popper has argued, are those statements reporting 'observations' that are commonly accepted, not having yet been challenged by any new discoveries or theoretical developments. To confirm this, all one has to do is review the controversies pertaining to the evidence offered in support of a new discovery or revolutionary theory, such as Galileo's telescopic observations, Darwin's evidence for evolution, or the current discussions of introspective reports pertaining to the mind-body problem.

Regarding Quine's second claim, "that all inculcation of meanings of words must rest ultimately on sensory evidence," again the issues are not as clear-cut as he would make it appear. If language

learning is not simply the behavioral reinforcement of the association of words with peripheral sensory stimuli under publically observable conditions, but depends also on innate endowments, then
this 'association' is located not at the periphery of our sense organs,
but someplace deeply embedded in the cerebral cortex. As previously stated, it is due to the British Empiricists that we even think
that knowledge begins with a peripheral stimulation of our senses;
yet as far as our *experience* is concerned, it is not peripheral stimulation that we observe, but the objects and occurrences in the
external world. Yet Quine draws what seems to me a most unlikely
conclusion: "What to count as observation now can be settled in
terms of the stimulation of sensory receptors, let consciousness fall
where it may."[11] But we do not observe light waves or photons
impinging on our eyes, or hear sound waves impacting on our ears;
we see colored objects and we hear sounds originating from certain
sources. Similarly, the other senses. We should not confuse the
causal explanation of perception with what we *consciously* perceive; moreover, the traditional view of perception as dependent
upon an image conveyed from the object via the senses to the brain
is also misleading. While the peripheral stimuli of our senses contain an encoded message from the object, it is not primarily our
sense organs, but our brains, that process this information enabling
the *conscious* organism to be aware of the external physical world
as experienced. The neural system is no more a conveyor of *sensory*
stimuli than a record player or television set. Where or how in the
brain the ultimate "transduction" or "transformation" of the chemical–electrical stimuli occurs is still a mystery—assuming the question is posed properly.

In fairness to Quine, it should be added that the indicated points
of disagreement are mainly a matter of emphasis, since he would
now accept much of what has been said. While stressing the importance of "sensory evidence" as conveyed in "observation sentences"
for science and for the "inculcation of meanings of words," he
repeatedly states that as language has grown more theoretical and
scientific theories more complex, the sensory or empirical contacts
have become progressively more attenuated.

> Science is a ponderous linguistic structure, fabricated of theoretical
> terms linked by fabricated hypotheses, and keyed to observable
> events here and there. Indirectly, via this labyrinthine superstructure,
> the scientist predicts future observations on the basis of past ones;
> and he may revise the superstructure when the predictions fail.[12]

Still, rather than considering *al experience* framework dependent, he insists that our theories are "keyed to observable events," and when predictions fail, it is the "superstructure" that needs revision. Yet the Copernican Revolution not only falsified the Ptolemaic superstructure, it emended our ordinary observations. If, as is now claimed, even "observation sentences" are theoretical, then this "keying" is a contingent and relative matter, for even "observable events" (such as the rising and setting of the sun) may have to be reinterpreted.

In any case, this preoccupation with language learning and observation sentences explains what must strike most people as Quine's curious approach to scientific knowledge, in terms of how scientists have learned their theoretical language and how this language is connected with observations. As Gibson states:

> *Quine's philosophy is a systematic attempt to answer from a uniquely empiricistic point of view, what he takes to be the central question of epistemology, namely, 'How do we acquire our theory of the world?'.* . . .[Furthermore] how we acquire our theory of the world. . .[is] the question of how we acquire our *theoretical talk* about the world. . . .[13]

Among the central inquiries of epistemology are questions such as how scientists have developed their theories and why these theories have proven so successful, but these investigations are quite different from the question of how scientists have acquired their "theoretical talk." I believe that any physicist, be it Newton, Einstein, Murray Gell–Mann or Sheldon Glashow, would be perplexed to be told that how he arrived at his theoretical discoveries is a question essentially of how he came to *talk* about the world as he did.

Could we equate Newton's derivation of the universal law of gravitation from Kepler's third law, Einstein's acceptance of the constant velocity of light and consequent revision of the concepts of space, time, and mass in the special theory, Gell–Mann's deduction of the theory of quarks based on the properties and interactions of subnuclear particles in conjunction with the laws of symmetry, and Glashow's and Georgi's (colleagues of Quine's at Harvard) formulation of their "Grand Unified Theory (GUT), with learning a theoretical language? Yet Quine claims that the main problem of epistemology is to account for the evidential basis of scientific theories in terms of how the scientist learns his theoreti-

cal language. As he says, "the scientist himself can make no sense of
the language of scientific theory beyond what goes into his learning
of it. The paths of language learning, which lead from observation
sentences to theoretical sentences, are the only connection there is
between observation and theory."[14]

Again, however, this manner of construing the problem in terms
of how scientists learn the language of scientific theory is curious.
Quine seems to think that learning a science consists of a linguistic
progression from observation to theoretical sentences. But for the
scientist this would be an artificial distinction since he learns the
laws, concepts, and formalism of a science by mastering the total
theory, which is accomplished in a number of ways: studying previous and current theoretical developments, doing experiments,
solving typical physical–mathematical problems, reflecting on
state-of-the-art issues, and so forth. None of this resembles "paths
of language learning." It is true that one begins with simpler and
then progresses to more complex theories, but this can hardly be
equated with advancing from observation sentences to theoretical
ones.

Unlike philosophers of science, who analyze historical scientific
discoveries and theoretical developments to gain some insight into
"The Nature of Natural Knowledge," Quine adopts the "genetic
approach" of studying how the theoretical language of science is
learned.

> We see, then, a strategy for investigating the relation of evidential
> support, between observation and scientific theory. We can adopt a
> genetic approach, studying how theoretical language is learned. For
> the evidential relation is virtually enacted, it would seem, in the
> learning. This genetic strategy is attractive because the learning of
> language goes on in the world and is open to scientific study. *It is a
> strategy for the scientific study of scientific method and evidence.*
> We have a good reason to regard the theory of language as vital to the
> theory of knowledge.[15]

But surely this "genetic strategy" is as artificial or as miscon-
ceived as Carnap's earlier program of "logial reconstruction,"
though for different reasons. To understand science we must learn
how scientists think, how they arrive at their discoveries, how they
design experiments, how they establish mathematical correlations
among phenomena, how they construct theories and equations to
explain the experimental results, and how they derive and test

predictions—not how they learn their theoretical talk! If Carnap's approach could appear to Quine as "make-believe," then Quine's own strategy seems to be sheer pretense. Could there be two things more dissimilar than the manner in which one learns a language and the method by which scientists resolve their theoretical problems? That the theory of language is "vital" to the theory of knowledge, especially scientific knowlede, is true in the trivial sense that scientists must have learned a language before they can begin to study any science, but to claim that the genetic approach to language learning provides a strategy for "the study of scientific method of evidence" is, to use Quine's term, "extravagantly perverse." Moreover, the theoretical language of the physical sciences is the symbolism of mathematics which is not learned the way one learns a natural language. Although his systematic approach to problems sets him apart from analytic philosophers, Quine shares with that school a tendency to view all problems as linguistic.

Turning now to more general considerations of language, it *is* incontestable that natural languages function as the common universe of discourse for a linguistic community. The means by which we focus our perceptions, shape our ideas, formulate our queries, and communicate our knowledge and beliefs, language provides our categories of recognition and classification, the basis of inductive inferences and causal explanations, along with the network of logical constants and conceptual interrelations that underlie our rational inferences and deductions. While individuals possess these cognitive abilities, it is our commonly shared linguistic framework that makes the acquisition and communication of them possible. Within this matrix, words acquire their more or less distinct and precise designations and connotations, while phrases and sentences attain their contextual meanings derived from the individual words as used within a particular linguistic context. For in spite of all the controversy about the "meaning of meaning," it is a fact that we do understand some words and sentences and not others, and that this is best rendered by saying that we do or do not understand their meanings.

As usually maintained, these word-meanings are learned by deliberate ostensive inculcation or reinforcement and casual imbibing, and are used initially as one word sentences. As more words are learned, they are arranged in strings or phrases, probably motivated by an innate grammar that helps shape correct word order. These initial phrases are succeeded by more extended and complex grammatical units with more diverse uses, such as affirmation and denial, conditionals, and so on. Eventually, contextual meanings and dictionary definitions supplant the earlier techniques of ostensive or casual word learning, especially with the progression of one's education. As scholars, we seem to forget that much of what we have learned beyond the common residue of meanings has been acquired from books and lectures.

In some such manner we acquire a web of knowledge and beliefs lodged in our systems of discourse. The web and network similes used so often to characterize language are meant to convey the interconnectedness of the strands of word–meanings inherent in language; the systematic interrelations that guide our discourse and underlie our inferential reasoning. Though there have been considerable misgivings regarding the analytic–synthetic distinction since Quine's classic article challenging the traditional "cleavage," there can be little doubt that a distinction can be made in most cases. Given our normal linguistic conventions, there is a marked difference in type of meaning, as well as in justification or testability, between such assertions as 'Paul is tall' and 'Paul is a male,' or 'this apple is red' and 'this apple is a fruit,' or 'what goes up must come down' and 'whatever is red is colored.' As Kant maintained, the first assertion in the three groups of sentences can be sensibly denied because the predicate is not 'contained in,' 'implied by,' or 'part of' the meaning of the subject term, whereas this is not true of the second assertion in the three sets of sentences.

Ignoring for the present Kant's dubious category of "synthetic *a priori* judgments," it seems to me that the examples he gives, as well as his justification for distinguishing between analytic and synthetic judgments, are correct. For instance, he distinguished between the judgments "All bodies are extended" and "All bodies have weight," declaring the first analytic and the second synthetic. His reason is that the former is "merely *explicative*" because the predicate has not "amplified in the least my concept of body, but. . .only analyzed it, as extension was really thought to belong to that concept before the judgment was made, though it was not expressed."[16] In contrast, he states that "All bodies have weight" is "*expansive*" or synthetic because it "contains in its predicate something not actually thought in the universal concept of body; it amplifies my knowledge by adding something to my concept. . . ." (p. 14). The "Common Principle" or criterion for distinguishing between the two judgments is the law of contradiction because negating an analytical statement consists of denying that the predicate is attributable to the subject term, *even though it constitutes part of its meaning,* which is contradictory. On the other hand, since in synthetic judgments the predicate is *not* part of the meaning of the subject term, and therefore may or may not be correctly predicated of it, there is no contradiction in negating such judgments, although the negation can result in a false statement. As Kant clearly states, "the predicate of an affirmative analytical judgment is already contained in the concept of the subject, of which it cannot be denied without contradiction" (p. 14).

It seems to me that Kant's distinction and criterion are straightforward and unobjectionable. Although the division depends upon the established conventional meanings or definitions at a particular time—a fact that we are more aware of today—this does not invalidate Kant's position. That *within* Aristotle's system the statement, "all [material] bodies have weight," is analytic has not been falsified by the fact that *today* we understand that a body has weight only within a gravitational field. As our linguistic conventions change, so do the lines of demarcation, but this does not refute the distinction. To use a well-worn example, at one time the statement "all whales are fish" would have been analytical, while today, given a more exact definition (though based on empirical evidence), the above statement is false, having been superseded by "all whales are mammals." It is a fact of language that there are certain attributes implied by the meanings of terms, so that to deny these attributes *is* to contradict their conventional meanings: a prime number is *not* odd; an electron is *not* a negatively charged particle; whatever is spherical is *not* circular; a brother is *not* a male sibling, and so on. As Kant correctly maintained, these statements, *given their conventional meanings,* cannot be negated or denied without contradiction, while the following can be: 'roses are red,' 'apples are sweet,' 'stones are heavy,' and 'gold is a yellow metal.'

I purposefully introduced the last example because, for Kant, it was analytical, given the conventions of his day, even though the statement is empirical. As he says, to know that gold is a yellow metal "I require no experience beyond my concept of gold as a yellow metal. It is, in fact, the very concept, and I need only analyze it without looking beyond it" (pp. 14–15). That is, even though the concept or definition of gold as a yellow metal is based on empirical evidence, *given the conventions of language in Kant's day,* the statement was analytical because 'being a yellow metal' is what was meant by 'gold.' This is no longer true because we now countenance white or pink gold, and know that there are yellow metals (such as "fool's gold") that have the appearance of gold but not its atomic structure. Kripke considers Kant's example of gold "an extraordinary one, because it's something that I think can turn out to be false."[17] Later he adds that "whatever necessity" Kant's example has "is established by scientific investigation; it is thus far from analytic in any sense" (f.n., p. 123). But according to my interpretation of analyticity, a statement can be analytic or not, depending upon the accepted linguistic conventions, though these conventions are based on empirical discoveries, as the examples of gold and whales illustrate. We have learned the meanings of many terms, such as 'hadron,' 'quark,' 'superego,' and 'four-dimensional manifold of events,' that are definitive of certain entities or states of affairs we have not ourselves experienced. Such statements, for us, are analytical.

Quine's objection to the analytic–synthetic "cleavage," as described by Kant, is that the criterion of self–contradiction "has small explanatory value" because it "stands in exactly the same need of clarification as does the notion of anlyticity itself," while the latter "appeals to a notion of containment which is left at a metaphorical level."[18] I do not consider either of these objections sound. If, as Kant maintained, by negation one denies an attribute to a concept that is part of the concept's meaning, this is contradictory: for example, 'copper is *not* a metal' or 'a triangle is *not* a three-sided figure.' Moreover, the meaning of "containment" is equally clear, namely, that if an attribute is contained in the connotation of a term, then it constitutes part of that term's meaning, as in red and being colored, father and having had a child, cat and being a mammal. I think it is Quine's aversion to meanings that really underlies his objection to the analytic-synthetic distinction because it does depend upon the meaning of the terms. While it may not make sense to say that words are contained in other words, it is appropriate to say so of meanings. Also, since Quine favors verbal dispositions over meanings, he would have to claim that the interrelations of meanings are "relations of interdependence between verbal dispositions" because he believes that "[w]e must study language as a system of dispositions to verbal behaviour, and not just surface listlessly to the Sargasso Sea of mentalism."[19] One may counter that neither should we allow our richly endowed subjective life to be swept away by the galloping tides of Le Mont-Saint-Michel, in the form of a behavioristic physicalism.

To illustrate the complex network of meanings embedded in ordinary language and how our normal inferential reasoning depends upon this network, consider the simple assertion "my daughter is away at college." From this statement we infer that the person making it is a father or mother, that the individual referred to is a female in her late teens, that she graduated high school recently with a certain scholastic record, that the college is located at some distance, and that the daughter is no longer living at home. All this, with varying degrees of specificity, can be inferred just from our understanding of the statement. Or consider the word 'water.' From this common term we immediately infer a substance that (normally) is a drinkable, transparent liquid, an excellent solvent, a fluid that can be transformed into ice or vapour under certain temperature conditions, that composes the rivers, lakes, and oceans (as salt water) of the world, that accounts for the largest percentage of the substances making up our bodies, that can be used to quench fire, and that has a molecular structure of H_2O. For the chemist, the last statement alone is packed with information concerning molecular compounds, atomic structures, chemical bonds, molecular and atomic properties, chemical reactions, and the like. Along with the logical connectives, these interrelated meanings enable us to weave the strands of discourse into intricate patterns of thought.

These meanings, with their implications, are what one must learn in any discipline. If, for example, one understands the term 'Newtonian mechanics,' then a nexus of concepts or meanings is implied: the existence of space and time as absolute frames of the universe within which indivisible, indestructible mass–particles move with mechanical necessity due to the forces of impact, gravity, and inertia. In addition, one would think of the universal law of gravitation whose strength varies with the product of the masses and inversely with the square of the distance, a law that explains Kepler's three planetary laws but not the precession of the perihelion of mercury. Should the mention of gravity suggest Einstein's general theory of relativity, then one understands that gravitational and accelerated effects are equivalent, that an inertial state and free gravitational fall are equivalent, that space contracts, time dilates, and mass increases either as the velocity of the system increases or as a result of stronger gravitational fields, that matter is a "condensed" center of a gravitational field which itself is reducible to a "distortion" or "wrinkling" of the four–dimensional, space–time continuum, Riemannian in structure.

Similarly, if confronted with the phrase 'origin of quantum mechanics,' one would probably think of the ultraviolet catastrophe, Planck's explanation via the absorbtion and admission of energy in discrete quanta of integral units, Heisenberg's principle of indeterminacy, the wave–partical duality and paradoxical results of the double–slit experiments, Bohr's quantum jumps to explain the emission spectra of hydrogen, the statistical nature of quantum phenomena as described by Schrödinger's wave equation and psi function or Heisenberg's matrix mechanics, Bohr's principle of complementarity, and the Copenhagen Interpretation of quantum mechanics. All these word–meanings and phrases or sentences, the implications of which vary with the expertise of the person, comprise semantic links in a complicated, cognitive–linguistic framework. As Paul Churchland says,

> a sentence is always an integrated part of an entire system of sentences: a language. Any given sentence enjoys many relations with countless other sentences: it entails many sentences, is entailed by many others, is consistent with some, is inconsistent with others, provides confirming evidence for yet others, and so forth. And speakers who use that sentence within that langugae draw inferences in accordance with those relations. Evidently, each sentence (or each set of equivalent sentences) enjoys a unique pattern of such entailment relations: it plays a distinct inferential role in a complex linguistic economy.[20]

Upon encountering something that jars or severs these semantic links, the individual infers immediately that something is wrong. Thus if someone were to say that quantum mechanics is based on continuous, determinate subatomic processes, and that relativity theory relativizes the laws of nature while establishing the independence of measurements of space, time, mass, and simultaneity from any coordinate system, we would conclude the speaker had fundamentally misunderstood both theories. Moreover, as indicated in the previous chapter, it is these theoretical implications that induce scientists to infer the existence of new physical entities, properties, or states, and derive predictions that are essential to the testing of any theory.

Again, what enables us to follow the argument and draw various inferences or conclusions (depending upon our individual understanding of what has been said), is that the words, phrases, and sentences carry various meanings. In my opinion, this interpretation, utilizing cognitive meanings (whatever their dependence on neurological processes) and their implicative interrelations, explains much more than Quine's description in terms of a "network of associations of words with words," or "of sentences associated with one another in multifarious ways. . . ." It is because Quine relies on such behavioristic principles as conditioning and association, rejecting the notion of meanings, that he is forced to admit that he cannot explain how we understand standing sentences (such as "sugar is sweet," whose truth is independent of the occasion of utterance): "I do not know how, in general, in terms of behavioral dispositions, to approximate to the notion of understanding at all, when the sentences understood are standing sentences."[21] However, this can be explained if we take into account the nexus of meanings. Again as Churchland states: "What begins as a set of stimulus–response patterns gets progressively articulated and refined as. . .the rich fabric of entailment relations characteristic of the theoretical framework as a whole."[22] It is this background web of meanings that enables us to explain knowledge and thought, the latter being essentially an inferential progression among meanings.

Moreover, Quine's description of the content of theories as "a fabric of sentences variously associated to one another and to nonverbal stimuli by the mechanism of conditional response,"[23] does not account for the tight network of meanings comprising theoretical frameworks. Are the interrelated concepts depicting subatomic particles, forces, and interactions in the Grand Unified Theory of quantum mechanics connected by the laws of association and conditioned response? I find Quine's account of the origin and structure of theories no more adequate than Hume's attempt to explain the perceptual content of our experience in terms of the "association of ideas," or of inductive inferences in terms of "custom" and "habit"! In light of what we know today about cognitive processes and structures, these radical empiricist assumptions seem completely inadequate, as I have argued throughout this book.

That words and phrases have a core conventional meaning is what enables them to be used in new ways and understood in novel contexts: for example, matter as a 'condensation' of a gravitational field and motion as a 'wrinkling' of the space–time continuum, or the emission of spectral radiation as due to electron 'jumps' among the orbital 'shells.' For this reason, I do not agree with the holist doctrine that the meanings of words are determined *wholly* by the context in which they are used. In fact, this thesis would be nonsensical if words shed *all* previous meaning in new contexts, since no meanings would be left to be newly established. Words have a core conventional meaning or dictionary definition which forms the basis of our understanding them when they are used in new or different contexts. This is especially true in literature where the figurative and/or affective meanings of terms are as important as their literal meanings, as in "Do not go gentle into that good night."

Recognizing that words have established conventional meanings that interconnect with the meanings of other terms also enables us to understand how transformations of theories involving alterations in word–meanings can occur without being incommensurable (as I argued in the previous chapter). It is just not the case that *every* attribute in the cluster of meanings of a term, along with its implicative connections, change when one theory, such as relativity theory, exacts a revision in our understanding of a previous theory, such as Newtonian mechanics. As they occur in relativity theory, the terms 'space,' 'time,' 'simultaneity,' and 'mass' acquire a *revised* meaning from that of classical mechanics, but not an *entirely* different meaning. For example, Einsteinian space, though a variable and relational dimension as measured from various non–stationary coordinate systems, in contrast with Newton's absolute space, still signifies the spatial separation or distance among events; similarly, although time, in relativity theory, ceases to be a 'uniform flow' independent of velocity or observers, it still designates the duration of an event or causal succession of events. In addition, while simultaneity and mass cease to be invariant, intrinsic properties of events or objects, they are precise designations or magnitudes maintaining a core of their traditional meaning as measured from a particular coordinate system: that is, mass retains the essential traditional meaning of resistance to motion, and simultaneity still means occuring at the same moment, as measured from *some* coordinate system. It is because of this that scientists refer to *relativistic* time, space, and mass, rather than use entirely new designations.

To illustrate more concretely, before the development of modern science people generally thought that qualities such as color, heat, and solidity were independently existing, intrinsic properties of all physical bodies; however, now that we realize that such qualities can only occur within a context where an organism with certain senses is a necessary component, we do not deny that our terms for colors, heat, and solidity have some continuity with the older terms. Or consider the continuous use but various modifications of the term 'atom' from Democritus, Newton, Laviosier, Rutherford, and Bohr to the present. When the meaning of a rival concept *is* incompatible and discontinuous with an older one, it *does* replace it, as occurred when 'oxygen' replaced 'phlogiston,' 'kinetic motion' replaced 'caloric fluid,' and 'field' replaced 'ether.' But the *replacement* of word–meanings is different from their gradual modification.

That our understanding and use of words is dependent upon an underlying cognitive matrix of meanings is true not only of verbal symbols, but is characteristic of mathematics as well. Consider all the meanings, relations, operations, and functions inherent in the number system. While originally, mathematical functions and theorems were derived from practical computations and measurements, after Euclid's axiomatization of geometry, mathematics was freed from all empirical confirmation, its valdity attributed to the self–evidence of the original definitions, postulates, and axioms, the deductive implication of the theorems from the axioms, and the consistency of the proofs.[24] Once arithmetic and geometry were accepted as formal calculi, mathematics developed by generalization, induction, and analysis into more abstract and complex systems: algebra, trigonometry, analytic geometry, differential and integral calculus, differential equations, topology, and so forth. At the same time, the number system was extended from natural numbers of positive integers to rational numbers (natural numbers plus zero and negative numbers), real numbers (irrational as well as rational numbers), imaginery numbers, and complex numbers. The seemingly endless possibility of mathematical developments derives from the multiplicity of meanings and implications *concealed* in the basic concepts and postulates, the generality and complexity of their interrelations, and the unlimited potential for enriching existing systems by extending and adding new notations and functions.

I believe this description of mathematics elicits the essential truth of Karl Popper's concept of a "3rd world," if we suppress its Platonic overtones. Popper distinguishes three independent, though interacting worlds: world 1, which contains the universe of physical entities, states, and processes comprising the macroscopic as well as the cosmic and subatomic domains; world 2, which includes all human mental or psychological states, both conscious and unconscious; and world 3, which comprises the artifacts of human creation, such as works of art, machines and technical devices of all kinds, intellectual creations, including scientific discoveries and theories, mathematical theorems, proofs, and equations, and philosohical and historical treatises.[25]

As products of the creative activity of human consciousness, world 3, whether objectified on canvas, in marble or bronze, on papyrus scrolls, parchment, or paper, represents the cultural heritage from which new interpretations, insights, and creations are derived. Once instantiated in paint or bronze, or printed in a musical score or manuscript, these human creations can be studied, analyzed, criticized, and reinterpreted by successive generations of artists, critics, and scholars. Though originated by human minds, they attain, after their creation, an independent, concrete existence as works of art, instruments, and books, which is why Popper accords them the separate status of world 3 objects. As such, they become the subject for further investigation by other minds, and thus contribute to the creation of more world 3 objects.

Given the medium and form of their creation, the materials of art, the alphabet of written language, and the notation of music, mathematics, and physics, there are embedded in the created works meanings and significances as rich and inexhaustible as the genius of their authors. This is why great creations are called universal, and can be reinterpreted by successive generations. Their significance is timeless. Inherent in the medium, then, is the potentiality for renewed exploration, discovery, and creation: the possibility of evoking representational images and abstract forms inherent in paints, bronze, and marble; the multiplicity of musical tones, variations, and melodies intrinsic to sounds and their composition; the potentiality for poetic rhyme and symbolism, literary themes, descriptive narration, and character development latent in language; the formal properties, relations, and functions concealed in the number system; the capacity of mathematical notation and operations to be used by scientists to depict physical properties and their causal relations, and so forth. While world 3 is normally instantiated in world 1 objects, as products of world 2 conscious beings they are endless sources of insight, inspiration, and reinterpretation.

The manner in which discoverable properties and functions are *concealed* in mathematical figures or concepts as abstracted from the real world can be illustrated clearly from standard functions found in any textbook of plane trigonometry. This is the branch of mathematics that deals primarily with six ratios, called the trigonometric functions (e.g., sine, cosine, and tangent) used in solving geometrical problems. Amazingly, the wealth of this subject derives from the ratios of just three lines of variable lengths conjoined so as to form variously shaped triangles with three interior angles. Since the two sides and the included angle suffice to fix the triangle's size and shape, the ratios of these sides for a particular angle determines the trigonometric functions (e.g., for a right triangle, sine $30° = \frac{\text{opposite}}{\text{hypotenuse}} = \frac{1}{2}$) which can be used to compute the length of the third side and the size of the remaining angles of the triangle. By abstracting and conjoining three straight lines, therefore, we can construct variously shaped triangles (equilateral, isoscoles, right, obtuse, and scalene) which have *embedded in their configurations* certain necessary ratios dependent on the lengths of the sides and the degrees of the enclosed angles. These ratios are the basis of the discovery of the trigonometric functions, identities, and equations that are *implicit in the constraints imposed by the particular triangular configuration.*

Analogous to mathematics as an abstract formal discipline, logic attempts to isolate from the wealth of ordinary propositions and arguments their logical form, along with the principles governing valid inferences. Aristotle began this discipline by reducing logical demonstrations mainly to class inclusion or exclusion relations, instituting the farsighted convention of using letters to stand for the verbally designated classes (thus, for the statement "All men are mortal," he substituted "All S is P"), specifying the kinds of valid and invalid arguments illustrated by these demonstrations. Aristotle's system of logic remained essentially unchanged until the middle of the nineteenth and early twentieth centuries when George Boole, Gottlob Frege, and Bertrand Russell laid the foundations of modern symbolic or mathematical logic.

Russell, especially, introduced new symbols, notations, and logical functions for analyzing propositions other than class inclusion (namely, predication and relations), for symbolizing more clearly the logical difference between existential and universal propositions, and for assigning truth functions to such logical connectives as conjunctions ($\wedge$), disjunctions ($/$), alternates (v), conditionals ($\supset$), and biconditionals ($\equiv$). Following the lead of mathematics, where alphabetical letters as variables are used in place of specific quantities, in logic small letters (x,y) stand for objects and capital letters (F, G) for the various functions and attributions, so that the statement "there exist some black swans" can be put in the following logical form: $(\exists x)(S_x \cdot B_x)$; namely, there is at least one object that is swanlike and black. Thus logic abstracts from the diversity of linguistic forms certain basic relations and functions, substitutes logical symbols for the linguistic conventions, and then specifies or develops proofs for the logical inferences that can validly be derived from sets of sentences. In this way, a logical matrix can be discerned among the great variety of linguistic conventions, inferences, and arguments.

As uninterpreted formal calculi, neither mathematics nor logic conveys any empirical information or knowledge, the latter being the function of natural and specialized languages, as well as the interpreted mathematical formalism of science. With the advent of modern science with its concerted search for lawful regularities among phenomena and verifiable causes and explanations, using increasingly refined experimental methods, the somewhat superficial representation of the world displayed in natural languages has been considerably emended. But while scientific developments with their technological consequences have been the major influence in modern times on our changing world view, it was the philosophers of the seventeenth and eighteenth centuries who were in the vanguard of the effort to interpret and assimilate the dislocating effects of these developments. Descartes was one of the founders of the mechanistic conception of the universe, while Locke, probably as a result of his close association with Boyle (whom he aided in his experiments at Oxford) and acquaintance with Newton, made the most deliberate attempt, from an empiricist orientation, to come to terms with the revolutionary scientific developments occurring at the time. Unlike Berkeley, who for religious reasons rejected the whole framework of Newtonian mechanics as consisting of meaningless abstract ideas, and Hume who disclaimed ever being able to discover the "hidden powers" and "insensible causes" of phenomena, Locke attempted to reconstruct our traditional conception of knowledge to accommodate the novel causal explanation of perception and the renewed interpretation of matter as atomic or corpuscular. Following Galileo and Boyle, he distinguished between "qualities" as physical properties and powers inherent in objects, and "sensory qualities" in our minds. Thus "secondary qualities," which cause sensations in us by affecting our senses, are "powers" inherent in objects due to the configuration and motion of the "insensible particles" (or atoms) which possess the "primary qualities" of size, shape, position, and motion. In

addition, objects have a third kind of power (analogous to secondary qualities) to affect changes in other objects (as when a fire
melts wax). Although we only experience sensations or ideas in our
minds due to the affects of insensible particles on our sense organs,
like his contemporaries Boyle and Newton, but unlike succeeding
philosophers such as Berkeley, Hume, and Kant, Locke maintained
that we *could* infer, to some extent, the real or primary qualities of
(atomic) matter.

This belief that we could infer, either directly or indirectly on the
basis of experimental evidence, something about the microstructure of matter and, therefore, explain the behavior of mac-
rophenomena (such as the gas laws and heat by the kinetic–
corpuscular hypothesis) was retained by the chemists of the eighteenth century, whose scientific principles and experimental discoveries laid the foundation of modern chemistry. Rather than
following the example of the scientists, however, the major philosophical tradition unfortunately took its cue from Berkeley and
Hume, which led to a successively more subjectivistic or idealistic
interpretation of knowledge, culminating in Kant's monumental
"transcendental" or "critical idealism."

> By *transcendental idealism* I mean the doctrine that appearances are
> to be regarded as being, one and all, representations only, not things
> in themselves, and that time and space are therefore only sensible
> forms of our intuition, not determinations given as existing by them-
> selves, nor conditions of objects viewed as things in themselves.[26]

Like his predecessors, Kant assumed we could have no knowledge extending beyond experience, the latter interpreted as sensory observation, not experimental inquiry: "Many forces in nature,
which manifest their existence through certain effects, remain for
us inscrutable; for we cannot track them sufficiently far by observation" (p. 514). We can "know" objects only as they appear to our
senses, not as they exist in themselves apart from our experience of
them. Our faculty of understanding, whose crucial function is to
provide judgments based on its inherent, invariant categories, is
limited in its application to the sensuous manifold as derived from
our sensory intuitions. This predetermined congruence between
the faculty of understanding and the sensory forms of intuition
provides objective or intersubjective knowledge of the phenomenal world: that is, the world of things in themselves or noumena *as*

they appear to man. Being thus limited to the *sensory* manifold, the categories cannot be applied to the *intelligible* world of noumena, reality as it could be known by some "non-sensory," "intellectual intuition."

> Doubtless, indeed, there are intelligible entities corresponding to the sensible entities; there may also be intelligible entities to which our sensible faculty of intuition has no relation whatsoever [i.e., entities that we, because of our natures as human beings, cannot sense]; but our concepts of understanding, being mere forms of thought for our sensible intuition, could not in the least apply to them. That, therefore, which we entitle 'noumenon' must be understood as being such only in a *negative* sense (p. 270; brackets added).

As the very concept of phenomena for Kant implies noumena, we must "think" of noumena as the non-sensory or *intelligible* complement of phenomena, though we cannot "know" them because our knowledge is limited to what can be sensed. "As setting limits to sensibility" the concept of noumena is "indispensable," but it is impossible to give the concept a positive content: "For we cannot in the least represent to ourselves the possibility of an understanding which should know its objects, not discursively through categories, but intuitively in a non-sensible [or intellectual] intuition" (pp. 272-273; brackets added). Although denying any possible knowledge of things in themselves or noumena, Kant nonetheless drops hints as to what he *thinks* such entities would be:

> ... if we could intuit ourselves and things *as they are,* we should see ourselves in a world of spiritual beings, our sole and true community...which has not begun through birth and will not cease through bodily death—both birth and death being mere appearances (p. 619).

Kant, of course, could not foresee the astonishing developments in knowledge that lay in the future, changes that would undermine the foundations of his "transcendental idealism." The introduction of non-Euclidean geometries, the theory of evolution, the tremendous advances in the neurosciences, and particularly the revolutionary developments in relativity theory and quantum mechanics negate his fundamental presupposition that our minds are so structured as to insure the *a priori* validity of Euclidean geometry and Newtonian mechanics *as regards nature, the phenomenal world, or reality as it appears to man.* As I emphasized in the previous chapter, Kant, no more than other savants of the seventeenth, eighteenth, and nineteenth centuries, could foretell the remarkable breakthroughs in knowledge in the twentieth century due mainly to the experimental technique of investigation. Because Kant, like Berkeley and Hume, limited possible knowledge to the domain of sensory observation, he did not anticipate that we could *experimentally* discover other "non-sensory" or "intelligible" properties of objects as they exist independent of our *observable* experience of them. Today we realize that while the categories of our natural languages portray the world as it appears under the usual conditions of experience, by experimenting with phenomena under modified conditions, we have been able to induce nature to reveal more of its non-sensory, intelligible, or intrinsic physical properties. Thus, Kant's inviolable distinction between the knowable world as it appears to man and the unknowable reality as it is in itself is no longer upheld or required by recent developments in science.

Furthermore, the whole basis of Kant's transcendental idealism crumbles if there are no "synthetic *a priori* judgments" at the foundation of knowledge, as he believed. For the purpose of his transcendental philosophy comprising space and time as *a priori* forms of intuition and the categories as inherent structures or concepts of the understanding, which therefore limit possible knowledge to appearances, was to explain how synthetic *a prior* judgments were possible in mathematics and physics. But it is precisely the unforeseen developments in these fields of knowledge that invalidate Kant's conception of synthetic judgments that could be known *a priori*. Instead, as I argued in the previous chapter, recent discoveries in science support the claim that we can experimentally discover *something* of the "insensible," "intelligible," or "inner natures" of things that justifies our belief in *a posteriori* necessary truths. These *necessary* truths discovered by *empirical* investigations constitute the implicative web of meanings in theories that underlies scientific inferences, explanations, and predictions—not transcendental mental structures! These, in turn, justify our belief in scientific progress, measured both in terms of increase in problem solving[27] and in the progressive unification of theories.

Because of the tremendous success of science in these areas during the twentieth century many scientists, despite historical warnings to the contrary, believe that a complete understanding of the world and man lies just beyond the present reach of scientific inquiry. However, this confidence is predicated on the belief that the knowledge acquired by experiental inquiry is sufficient for a total explanation of the universe. If this proves to be false, which I believe is possible, then there is a sense in which Kant's position is true after all—in that we may have extended the boundary between the knowable and the unknowable much farther than Kant would have expected, without having removed it entirely. But should this prove to be the case, rather than returning to a neo–Kantian transcendentalism, I believe the position of "contextual realism" that will be presented in the last chapter provides a better framework of interpretation.

Notes to Chapter V

1. Although this phrase has become commonplace, it was initially used by Kant in the *Critique of Pure Reason,* trans. by Norman Kemp Smith (New York: Humanities Press, 1933), p. 121.
2. Cf. W. V. Quine, "The Nature of Natural Knowledge," in Samuel Guttenplan (ed.), *Mind and Language* (Oxford: Clarendon Press, 1975), p. 69.
3. Cf. D. W. Hamlyn, *Experience and the Growth of Understanding* (London: Routledge & Kegan Paul, 1978), p. 90.
4. W. V. Quine, *Ontological Relativity and Other Essays* (New York: Columbia University Press, 1969), p. 26. Italics added.
5. Peter D. Eimas, "The Perception of Speech in Early Infancy," *Scientific American,* January, 1985, p. 46.
6. W. V. Quine, *Ontological Relativity, op. cit.,* p. 75. Brackets added.
7. Cf. *ibid.,* p. 89.
8. Paul M. Churchland, *Scientific Realism and the Plasticity of Mind* (Cambridge: Cambridge University Press, 1979), p. 139.
9. Paul Feyerabend, "Science without Experience," in *Challenges To Empiricism,* ed. by Harold Morick (Indianapolis: Hackett Publishing Co., 1980), p. 161.
10. C. I. Lewis, *Mind And the World Order* (New York: Dover Publications, 1929), p. 261.
11. W. V. Quine, *Ontological Relativity, op. cit.,* p. 84.
12. W. V. Quine, "The Nature of Natural Knowledge," *op. cit.,* pp. 71–72.
13. Roger Gibson, *The Philosophy of W. V. Quine* (Tampa: University Presses of Florida, 1982), pp. xviii–xix. Brackets added.
14. W. V. Quine, "The Nature of Natural Knowledge," *op. cit.,* p. 79.
15. *Ibid.,* pp. 74–75 (italics added).
16. Immanuel Kant, *Prolegomena to Any Future Metaphysics,* revised Carus trans., fifth ed. (Indianapolis: Bobbs–Merrill, 1950), p. 14. The immediately succeeding quotations are from this work.
17. Saul A. Kripke, *Naming and Necessity* (Cambridge: Harvard University Press, 1972), p. 39.
18. W. V. Quine, "Two Dogmas of Empiricism," in *From A Logical Point Of View* (Cambridge: Harvard University Press, 1953), pp. 20–21.
19. W. V. Quine, "The Nature of Natural Knowledge," *op. cit.,* p. 91.
20. Paul Churchland, *Matter and Consciousness* (Cambridge: MIT Press, 1984), p. 21.

21. W. V. Quine, "Mind and Verbal Dispositions," in Samuel Guttenplan (ed.), *op. cit.,* p. 89.
22. Paul Churchland, *Scientific Realism and the Plasticity of Mind, op. cit.,* p. 29.
23. W. V. Quine, *Word and Object, op. cit.,* p. 11.
24. Cf. Richard H. Schlagel, *From Myth to the Modern Mind,* Vol. I, *Animism to Archimedes* (Bern: Peter Lang, 1985), pp. 252–258.
25. Cf. Karl R. Popper, *Objective Knowledge* (Oxford: Clarendon Press, 1972), pp. 153–155; also Karl R. Popper and John C. Eccles, *The Self and Its Brain* (New York: Springer International, 1977), pp. 41–43.
26. Immanuel Kant, *Critique of Pure Reason, op. cit.,* p. 345. The immediately succeeding quotations are from this work.
27. Cf. Larry Laudan, *Progress and Its Problems* (Berkeley: University of California Press, 1977), p. 5.

CHAPTER VI

THE MEANING OF TRUTH AND THE CORRESPONDENCE CRITERION

MY intent in the preceding chapters was to interpret the implications of recent developments in neurophysiology, physics, cognitive psychology, linguistics, and the philosophy of language for understanding the preconditions, origin, and nature of knowledge. Based on this assessment, it was concluded that our acquisition of knowledge can be understood best if we interpret it as fostered by certain innate cognitive and linguistic endowments founded on inherited neurological capacities, rather than in terms, solely, of the empiricist principles of reinforcement, conditioning, and association. But though our intellectual abilities are dependent upon intricate processing functions of the brain, we must also take into account the conscious complements of these activities, such as simple awareness and intention, along with concepts and meanings. There is no proof as yet that man is merely a biological robot! The implicative nature of knowledge was also emphasized as manifested in the network of meanings embedded in our various cognitive–linguistic frameworks. But while all knowledge is framework dependent, I argued, in opposition to Hume, that the systematic entailment of scientific knowledge consists largely of *necessary empirical connections* based on our current understanding of the intrinsic structures and properties of physical objects, chemical reactions, and physiological processes disclosed by experimentation.

Turning now from a *descriptive* account of knowledge, the following chapters will be concerned with the problem of the *justification* of knowledge: that is, with the meaning and criteria of truth, along with the nature of reality implied by these criteria. We shall, thereby, avoid the error of confusing a causal explanation with a justification of knowledge, what Rorty refers to as "the basic confusion contained in the idea of a 'theory of knowledge.'"[1] A similar distinction is stressed by Sellars:

> The essential point is that in characterizing an episode or a state as that of *knowing,* we are not giving an empirical description of that episode or state: we are placing it in the logical space of reasons, of justifying and being able to justify what one says.[2]

The claim to know something requires stating some fact and being able to justify that it is a fact. Thus if someone claims to know the date of an historical event such as the Battle of Hastings, the physical properties of an element such as krypton, or that it is snowing outside, that person must be able to express the knowledge claim in the form of meaningful statements that designate and delineate the fact or state of affairs. The expressed *content* of knowledge is conveyed by the statement and often is referred to as a proposition (not a reified entity, but the understood meaning), yet what the knowledge claim pertains to is not the proposition but what the proposition depicts.[3] If I know that the Battle of Hastings occurred in 1066, that krypton is a colorless inert gas with an atomic number of 36 and atomic weight of 83.80, or that a snowstorm is raging in the immediate vicinity, then my knowledge is about these facts, entities, or events. The statement is what we usually designate as true or false depending upon whether what it represents, the fact or state of affairs, is in accordance with this representation or not, while the fact or state of affairs is neither true nor false, but either is or is not the case.

Thus claiming to know something places on the individual a certain obligation to be able to articulate and justify what is being claimed, a type of knowledge that is usually referred to as sentential or propositional. The various modalities of the claim, such as knowing, believing, or doubting have, owing to Russell, been described as "propositional attitudes." Also, there is some consensus among philosophers that propositional knowledge can be equated with justified true belief. In this view, an individual (x) is justified in claiming to possess certain knowledge expressed by the statement (s), if and only if

1. x *believes* that s is true,
2. s *is* true (because what it designates is the case),
3. x is *justified* or *warranted* in believing s.[4]

To know that a statement is true, then, entails that one believes the statement to be true, that the statement is in fact true (since one could not claim to truly know something if his claim were actually false), and that one be accountable for his claim to know, in being able to provide good reasons or adequate justifications for it (since one would not credit someone with knowing something, though his claim was true, if the grounds of his knowing it were unsound). But while it is the statement or belief that we normally designate as true or false, our concern, as we shall see, is not with the statement *per se*, but with what the statement asserts: what it is about. If, then, knowing that something is the case consists of assenting to what is true on justifiable grounds, two fundamental questions are raised: what do we *mean* by truth?, and how do we a4justify a statement as being true?

The Meaning of Truth: Taking a cue from Alfred Tarski, we will begin with Aristotle's definition of truth:

> . . . we define what the true and the false are. To say of what is that it is not, or of what is not that it is, is false, while to say of what is that it is, and of what is not that it is not, is true; so that he who says of anything that it is, or that it is not, will say what it true or what is false. . . .[5]

In this explication of truth, Aristotle, like Sellars in the earlier quotation, states correctly that truth pertains to assertions, to what one "says." Until something has been affirmed, some truth claim made, the question of whether what has been affirmed "is or is not," will not arise. Thus, unlike so many others (Hegel's assertion, for example, that "the true is the whole"), Aristotle did not confuse the issue of truth with being or reality. That something is or is not the case, does or does not display a certain feature, exists or does not exist pertains to the actuality, existence, or reality of a state of affairs. It refers to an actual condition of the world which is there independently of whether we assert anything about it or not, attesting to the instinctive realism or "animal faith" in an independent reality that underlies any conception of objective knowledge or truth. On the other hand, judgments, beliefs, statements, or assertions are true or false depending upon whether what they state, designate, or represent is the case or not.

Unfortunately, however, the problem is more complex than implied by Aristotle's definition. Under the most straightforward interpretation, Aristotle would appear to be making the common sense assumption that the world is, just as it is given to us in experience— that the world as 'we' know it is not affected by any limiting or distorting experiential conditions or conceptual misinterpretations, but is given as an unmodified, antecedent reality. Thus if we say of "what is" *that it is,* then the statement will be true; if we say otherwise, it will be false. But this reference to "what is" is ambiguous because it does not distinguish between what *really* is an independent feature of the world and those features which, though seemingly objective, are actually a function of our experience (such as geocentrism and retrograde motion) or interpretation (as Aristotle's conviction that nature abhors a vacuum or Newton's belief in absolute space and time). In one sense, the reference to "what is" simply implies that given an assertion about something, the assertion will be true (or false) depending upon whether what it states is the case or not. However, there is a deeper question involved, namely, whether the assertion itself, *even when verified,* describes the world as it is, or whether it merely *conveys a confirmed misrepresentation* of what exists. As Churchland states, "we may be systematically *mis*perceiving reality in the first place."[6]

Thus two assumptions would seem to underlie the most obvious interpretation of Aristotle's definition: (1) that an independent world is directly presented to us in experience, and (2) that our conceptual–linguistic framework mirrors the world as it is. On these assumptions, "to say of what is that it is," is to assert the truth. *But how do we know that what we "say" truly represents the world as it is?* For example, in one sense the statement "the sun rises and sets in relation to a stationary earth" is true, but in a deeper sense it is not, because the rising and setting of the sun are only apparent phenomena due to the rotation of the earth on its axis. Similarly, we describe fire as painful, yet the assertion is false if it is taken to mean that the pain exists within the fire; but, as Locke asked, is there any real difference between the above assertion and stating that the sky is blue or that sugar in itself is sweet? Yet, as regards these latter examples, we do assume that *it is* the sky that is blue and the sugar *by itself* that is sweet! Although certain linguistic uses have a conventional sanction, how do we know that they represent a real state of affairs? This is the problem that the most obvious interpretation of Aristotle's definition does not take into account.

On the other hand, if we ignore the realistic assumptions of the rest of Aristotle's philosophy, we could suppose that he *was* aware of the problem and expressly formulated his definition of truth so as to take it into account. On this supposition we would conclude that Aristotle was not only aware of, but accepted the fact that all knowledge is framework dependent, hence that a direct or naive realism is false, and that "what is" could have no meaning apart from some linguistic formulation. That is, while an adherent of direct realism assumes that the world is presented to us at it is, truth consisting simply of so describing it (presuming one has the correct means of description), an adherent of the converse position that all knowledge is framework dependent, maintains that whatever we can *mean* by something "being what it is" depends ultimately on our conceptual–linguistic framework (although whether a *particular* assertion *within* that framework is true or false will depend upon how the world happens to be at hat moment), as Popper, [7] Quine, [8] Sellars, [9] and Feyerabend [10] maintain.

According to the latter position, since we can never know the world apart from some theoretical interpretation, all assertions of truth (or falsity) are made within the context of such interpretations, and thus are unavoidably relative to them. Hence any question as to what reality may be "in itself," apart from such theoretical frameworks, is unanswerable or fruitless, but not *meaningless.* Although this issue has often been confused, the claim that all knowledge is framework dependent does not deny that there is an unknowable reality apart from or beyond that recognized in any framework, but just the opposite. Thus Kant's reference to a domain of "things in themselves" apart from our experience was not meaningless, as positivists and ordinary language philosophers asserted, though his particular delineation of it may not have been the most acceptable. As Kant maintained, in contrast to Wittgenstein, if we are forced to acknowledge a bound to what we can know (or say), this implies the existence of something that we do not know (or cannot talk about). [11] It is only if we claimed to *know* this unknowable reality that we would be contradicting ourselves (Kant, for example, never claimed to "know" such a reality, since we lack any intuition of it, but he did endeavor to explain how we might "think" of it[12]). Acknowledging that there is more to reality "than is dreamt of" in any of our conceptual–linguistic frameworks is not only not contradictory or meaningless, but the only sensible position to hold, in my opinion.

However, the thesis that all knowledge is framework dependent does not preclude an independent world from affecting our truth claims—on the contrary, the world exerts its influence in two ways. First, as different in syntax as natural languages might in some cases be, it is unlikely that any language or conceptual system could be so arbitrary or fanciful that it would not reflect at least some objective features of the *experienced* world, however strange these features might appear from the standpoint of another linguistic system. For example, though the Hopi language does not conceptualize some actions in the way that Indo–European languages do, utilizing different linguistic conventions to express temporal occurrences, the Hopis are not unaware of such activities or of the difference between processes that have been completed and those that are either beginning or taking place.[13] While there are different ways of conceptualizing and expressing the same objective state of affairs, any conceptual–linguistic system must 'fit' nature, at least to the extent that it imposes some intelligible and practicable structure on it, enabling one to follow and anticipate the course of events, as well as discriminate among beneficial and harmful effects. Survival would not be possible otherwise.

The continuous testing and subsequent modification or rejection of theories, as Karl Popper has stressed,[14] supports the same conclusion. As different as were the cosmologies of Democritus, Plato, and Aristotle, each fitted some features of our experience of the world: that is, each was based on some recognizable abstractions or it would not have had (and for some philosophers continue to have) a certain plausibility. So, too, the Ptolemaic model of the universe, although subsequently replaced, would not have had such a long history if it had not agreed with, accounted for, and facilitated predictions of certain astronomical phenomena. Thus the model was not entirely misguided, but approximated the true structure only to a certain extent from a particular perspective, that of geocentrism with the apparent diurnal rotation of the "heavens" around a stationary, central earth.

As is well known, Copernicus rejected the geocentric view because it was too *ad hoc* and cumbersome, and did not provide as uniform or simple an explanation as the heliocentric model for such planetary phenomena as retrograde motion, variations of brightness, and slight alterations in orbital periods. But without the previous organization of astronomical observations into the two-sphere model of the universe, these more exact and obscure data would not have been recognized at all, or, had they been, would not have been accorded any significance.

An especially dramatic example of nature intruding into established scientific categories, and consequently forcing a revision of fundamental concepts, was the outcome of the Michelson–Morely experiments. At the end of the nineteenth century physicists continued to believe in the existence of an absolute space, and in an ether filling this space, which they considered necessary for the propagation of electromagnetic waves (since waves are only 'propagated' in a medium). In addition, they assumed that the Galilean principle of the addition of velocities would apply to light, so that as the earth revolved in its orbit through the ether, its motion would affect the velocity of light in such a way that the 20 miles per second velocity of the earth would have to be added to or subtracted from the 186,000 miles per second velocity of light, depending on their respective directions of motion. Michelson and Morely were incredulous to find, however, that the velocity of light remained unchanged under all relevant experimental conditions.

Einstein, who had already deduced this fact, along with the variation of spatial and temporal measurements among observers in relative motion, accounted for the "null results" of the Michelson–Morley experiments by reversing the fundamental presuppositions: instead of the velocity of light being variable, and spatial and temporal magnitudes being invariant among observers in relative motion, he accepted the constant velocity of light and deduced that spatial and temporal magnitudes must vary according to the velocity of the system in which they are measured, thereby revolutionizing Newtonian mechanics.[15] As the example illustrates, established concepts and theories function as cognitive frames for interpreting phenomena and making observations, as well as paradigms for guiding further research, deriving predictions, and testing for validity. In this way, the matching of theories with nature is constantly being challenged, so that neither theoretical interpretation nor nature alone is sufficient to account for the development of knowledge.

The second way nature imposes itself on our experience is in the verification of *specific* affirmations *within* the context of a particular conceptual–linguistic framework. Even though the designation and characterization of states of affairs depend upon some linguistic expression, this does not mean that at any one time the world will exhibit that particular state of affairs. For example, in English one says "the sky is blue," while in Russian one says the equivalent of "the sky blues," but regardless of the particular linguistic formulation, whether or not the sky appears the color of blue at any one time is independent of any assertions about it. The way the world is has something to do with our truth claims. Thus the fact that a state of affairs is *identified* by a verbal expression does not *determine* or *confirm* that such a state does in fact exist at any particular time— were this the case, we could create truths by linguistic fiat.[16]

Moreover, we should distinguish between two kinds of factual truths, those whose truth is (relatively) independent of the occasion on which they are asserted, and those whose truth depends specifically upon the occasion (Quine calls the latter "occasion sentences," and the former "eternal" or "standing sentences"). By the former I mean such statements as "the English alphabet consists of twenty-six letters," "Newton was born the year Galileo died," "neutral pions decay into two photons," "Thursday precedes Friday," and so forth. Although none of the assertions could be made prior to a knowledge of the existence or the occurrence of the facts they describe, still, their truth is not dependent upon the occasion or circumstance of a particular person making the assertion. In contrast, the truth of such statements as "John is speaking," "it is raining outside," "today is Thursday," or "I am tired" does depend upon the occasion on which they are asserted, or on who is making the assertion. Thus the fact that Newton was born the year Galileo died is true regardless of who asserts it (assuming the assertion is made after Galileo died and Newton was born). However, the truth of statements such as "it is raining outside" or "today is Thursday" is indeterminate until referred to a particular situation or context. To use Searle's term, the truth of the latter assertions is tied to specific "speech acts," while the truth of the former is independent of the instance or situation in which they are asserted. This was the point of Strawson's analysis and clarification of Russell's perplexity regarding such statements as "the present king of France is bald," uttered at a time when France was no longer a monarchy.[17] Such examples indicate the degree to which the truth value of synthetic statements depends upon the context, occasion, or circumstance in which they are asserted. On the one extreme, the truth of $A=A$ or $1+1=2$ is practically context free, while the assertion "I am tired" is completely indexically dependent.

Returning to our discussion, regardless of what Aristotle himself intended (although it is unlikely that he meant anything more than the simple realistic interpretation), the notion that all knowledge is framework dependent would seem to be the more plausible view from the contemporary standpoint. The merit of Tarski's well-known "semantic conception of truth," derived from Aristotle's definition, is that it explicitly formulates a conception of truth consistent with the notion that all knowledge is based on a framework of interpretation. Tarski accomplishes this by the simple expedient of using single quotation marks to distinguish between a statement and its referent.[18]

While normally the transparency of language in everyday discourse deludes us into thinking that such a distinction is unnecessary, the previous considerations along with the "linguistic relativity" thesis of Whorf and Quine discussed in Chapter III, should convince us that the situation is not so simple. If each linguistic framework conditions our experience and conception of the world, and if we can never know the world except by means of some such framework, then how can we know what the world would be like independent of any framework (or unconditioned)? This is Kant's problem transferred from a cognitive to a linguistic context. Also, the situation is analogous to the older paradox of perception: how can we catch a glimpse of what the world would be like—as it is independent of us—apart from glimpsing it? How can we conceive of what the world is like, apart from some conceptualization of it?

Tarski's semantic conception of truth concedes the notion of framework dependence, using verbal expressions as surrogates for states of affairs, distinguishing the former from the latter with single quotation marks. As his own example illustrates, the assertion 'snow is white' is true if and only if snow is white.[19] In this example, the statement "snow is white" occurs twice, first as a *verbal expression* designating a state of affairs, and second, as *substituting for* the state of affairs itself. This second use provides linguistic 'stand-ins,' so to speak, for the actual states of affairs or facts. The distinction is useful because there are things that can be said meaningfully of one use of the statement but not of the other: one can say of the statement 'snow is white' that it is written in English and that it consists of eleven letters, characteristics not attributable to snow; conversely, one can say that snow is white because it absorbs and reflects certain light waves which would make no sense if attributed to the written statement.

But the primary advantage of Tarski's formulation is that it takes into account the fact that states of affairs can, for all practical purposes, only be delineated and identified by means of linguistic expressions (as Wittgenstein suggested, try communicating to someone a designated fact, such as the color of a book, simply by pointing to it). Moreover this designative function is not limited to physical states of affairs, but applies to all kinds of theoretical facts, such as the fact that the sum of any two prime numbers is an even number, as well as to hypothetical situations, such as that in the next century, the science of genetic engineering will have progressed to the point where society can program the kinds of individuals it wants or needs. All such statements can be fitted into Tarski's general schema, '................' is true if and only if .... A further advantage of this formulation of truth, as Tarski indicated, is that it leaves open the question as to how or when the confirmation or disconfirmation takes place. What it asserts is that we use the term 'truth' (or conversely, 'falsity') to characterize certain statements, namely, those of which the states of affairs or facts conform (or do not conform) to what has been designated by the statement.

However, as Tarski's semantic conception of truth depends upon the distinction between a meta–language, in which the definition of truth is formulated, and an object–language, to which the concept of truth applies, and pertains in a rigorous way only to formalized, artificial languages, one wonders, as Max Black did, "how far the definition could be adapted to 'ordinary' English or any other 'natural' language. . .[and] how far the results illuminate 'the *philosophical* problem of truth.'"[20] The latter problem, according to Black, consists not only of explicating instances of truth (as 'snow is white' is true if and only if snow is white), but of providing a general description of the use and meaning of truth, and of the nature of the criteria used in establishing the truth of an assertion.[21] We shall begin, then, with whether there is an ordinary use and meaning of the term 'true.'

Although it occurred as long ago as 1950,[22] I shall draw on the debate between John Austin and Peter Strawson on "Truth" to illustrate the most salient features of our ordinary use and meaning of the terms 'true' and 'false,' when we indicate our agreement or disagreement with what someone said on a particular occasion. There are many other uses of the terms (and their equivalents), but the latter tends to be central. In what follows I make no pretense of describing all the subtlies of the debate or all the facets of each position. My intent is to extract what we mean by truth when used as indicated.

Austin: (1) There is a "primary" or "generic" use of such predicate adjectives as true and false in ordinary natural languages.[23] (2) The rejection of Ramsey's claim that true and false are not genuine predicates since they serve merely to add emphasis to a statement, maintaining instead that they are not redundant or "logically superfluous" (cf. p. 25). (3) True and false (and their equivalents) are genuinely predicated of statements or assertions "*as used by a certain person on a certain occasion*" (p. 20). (4) The purpose of predicating such adjectives to these statements is to affirm that a certain relation holds between the former and the world, namely, the relation of agreement or correspondence (cf. p. 21). Though he claims that it can be misleading, Austin's direct answer to the question, "When is a statement true?", is, "'When it corresponds to the facts'" (p. 21). He says "[t]here are numerous other adjectives which are in the same class as 'true' and 'false,' which are concerned, that is, with the relations between the words (as uttered with reference to an historic situation) and the world" (p. 28). (5) Whatever words are used in any language to refer to or describe the world is a matter of convention, but given these conventions, whether an agreement exists between the two is a question of fact or truth.

> When a statement is true, there is, *of course,* a state of affairs which makes it true and which is *toto mundo* distinct from the true statement about it: but equally of course, we can only *describe* that state of affairs *in words*....(p. 23).

Thus "the truth of statements remains...a matter...of the words used being the ones *conventionally appointed* for statements of the type to which that referred to belongs" (p. 25).

Strawson: Although Strawson agrees with Austin, as opposed to Ramsey, that the use of true and false are not redundant or logically superfluous, his essential thesis is what while certain conditions must be fulfilled for a statement to be true, Austin's "central mistake is to suppose that in using the word 'true' we are asserting such conditions to obtain."[24] Strawson denies that when we affirm a statement to be true we are asserting a relation of correspondence or agreement between the statement and the world, arriving at this conclusion by the following arguments. (1) He denies "that to say that a statement is true is to say that a certain speech–episode is related in a certain conventional way to something in the world exclusive of itself" (p. 32), which is his description of Austin's position. He concedes that we predicate truth or falsity of statements, but not as historic events or speech–episodes: it is what is *said* that is true, not the *act* of saying it. Like Austin, he distinguishes between the referring and the descriptive parts of statements: statements are "about" what they refer to and describe.

(2) His main objection to Austin's position (completely unfounded, as I shall try to show) is that Austin held that, in addition to statements being about the world in the above senses, there must be *something else* in the world, namely *facts,* to which the statement is *related,* if it is true:

> ... the demand that there should be such a relatum is logically absurd: a logically fundamental type–mistake. But the demand for something in the world *which makes the sttement true*...or *to which the statement corresponds when it is true,* is just this demand (p. 37).

This, however, is a mistaken interpretation, since Austin did not presuppose any additional relation of correspondence beyond the fact that what a statement is about must be the case for it to be true: that the truth of a statement consists of a relation of agreement between the statement and whatever there is in the world that the statement designates. The reason Strawson gives for claiming that Austin must assume there is something in the world that makes the statement true, in addition to what the statement is about, is that "while we certainly say that a statement corresponds to (fits, is borne out by, agrees with) the facts...we *never* say that a statement corresponds to the thing, person, etc., it is about" (p. 37). In terms of what he claims we can or cannot say in ordinary discourse, Strawson attributes to Austin a position that he did not hold, though Strawson is correct in criticizing him for not distinguishing between the term "fact" and such other terms as "situation," "event," and "state of affairs."

But though Austin was somewhat imprecise (ironically, considering his obsession with language) in his use of 'fact,' there is no justification for Strawson drawing the following conclusion: "The only plausible candidate for the position of what (in the world) makes the statement true is the fact it states; but the fact it states is not something in the world" (p. 37). I believe a careful reading of Austin indicates that rather than using the term 'fact' to cover something *not* in the world, he uses the term to designate, and thus as *equivalent to,* those states of affairs or conditions *in* the world that the statement is about, and on which its truth depends. Austin explicitly says that "speaking about 'the fact that' is a compendious way of speaking about a situation involving both words and the world" (p. 24). As Strawson maintains, "[f]acts are what statements (when true) state; they are not what statements are about" (p. 38). This is correct, and I do not believe Austin intended to claim otherwise.

(3) Strawson asserts that when Austin maintained that the relation between the words used by people to talk about the world is merely conventional, he also held that when we declare a statement to be true we either are "talking about the meanings of the words used by the speaker" or "saying that the speaker has used correctly the words he did use" (p. 44). Yet Austin denies this: "we never say 'The meanings (or sense) of the sentence (or of these words) is true" (p. 20). What Austin was claiming is that even when statements are true, "the correlation between the words (= sentences) and the...situation, event etc...is *absolutely and purely* conventional" (p. 24).

(4) Strawson then concludes, and this is the essential point, that the whole notion of a correspondence between a statement and anything else is a mistake:

> It is the misrepresentation of "correspondence between statement and fact" *as a relation, of any kind, between events or things or groups of things* that is the trouble. Correspondence theorists think of a statement as "describing that which makes it true" (fact, situation, state of affairs) in the way a descriptive predicate may be used to describe, or a referring expression to refer to, a thing (pp. 40-41).

In opposition to Strawson, I shall try to show that, as a matter of fact, our normal use of predicate adjectives such as true and false *is* precisely to confirm a relation between a statement and what it is about: that the whole question of truth would collapse if, when we assert that a statement is true, we were not also claiming that what it designates is the case.

(5) Finally, in addition to pointing out that there are numerous ways and contexts in which true and false are used, Strawson's main

point is that they normally are not applied to speech acts or episodes. Instead, he maintains that when we use true and false to characterize someone's assertion or statement, our usual intent is to show our "corroboration" or "agreement" with what has been said, not the saying of it. Thus he accuses Austin of confounding the query, "*When* do we use the word 'true'?", with the question, "*How* do we use the word 'true'?" (p. 44). Again, however, I think this is a false criticism. Not only was Austin maintaining that we use the terms 'true' or 'false' *when* we want to state, on the appropriate occasion, that what another person asserted was correct or incorrect, that it corresponded or did not correspond to the situation designated, he also intended this as an explanation of *how* we use the terms: namely, to indicate whether the state of affairs as designated is the case or not. Strawson's own analysis, that in such situations we use such terms to express our *agreement with what the speaker said,* not to state that certain truth conditions have or have not been fulfilled, is no more or less an answer to the *how* question than Austin's.

More to the point, however, is either Austin's or Strawson's position correct? In the usual circumstances when we use predicate adjectives such as true–false, agree–disagree, and yes–no, are we either predicating something of a statement or expressing our agreement with the person who made the assertion? Suppose, for example, someone says after a winter storm, "it is too slippery to drive," and another person replies, "that's true." Is the latter expression intended *primarily* to affirm the original statement or to express one's agreement with the speaker, or is it intended to confirm that the road conditions are as described? That is, when one expresses agreement with what a person said, there are three possible interpretations: (1) agreement with what the *speaker* said; (2) agreement with what the speaker *said*; and (3) agreement with *what* the speaker said. Strawson claims (erroneously in my opinion) that Austin supports the first interpretation, while he supports the second, but I think the third is the correct one.

The *primary* purpose of fact–stating discourse, while having as secondary effects the first two functions, is to state something (true) about something—usually some state or circumstance pertaining to the world. When we use such typical expressions as "you misspelled that word," "the meeting will be on Tuesday," "you are late for the appointment," "your car needs a tune-up," or "the fare has gone up," and someone says, "Yes, that's true," "I agree," or "you are right," the agreement is primarily with what the statement is about, and only secondarily with whomever makes it, or with whatever words are used in making it. As Warnock correctly states in his critique of the debate between Austin and Strawson:

> In saying to me truly, for instance, 'It's raining heavily', you have, *intending to comment on current weather conditions in our vicinity,* correctly produced exactly that utterance best adapted to our common language to this very purpose: you have not only successfully drawn my attention to the place and time intended, but have also characterized correctly the relevant aspect or feature of what is going on there and then.[25]

Having fulfilled these conditions, the second person, in responding affirmatively, is primarily expressing agreement not with the speaker or with the statement made by the speaker, but with "the relevant aspect or feature of what is going on there and then." This is indicated particularly in a negative response. If someone says, "It is raining," and another responds, "No it's not," the "it" referred to in the denial does not represent either the person or his assertion, but the condition of the weather.

I am not denying, of course, that the predicate adjectives also have the secondary functions, but when we want to call attention primarily to these secondary functions, we use different locations. For example, if we wish to indicate agreement with the *statement* made by the speaker, we are apt to use expressions such as, "your choice of words is perfect," "I agree with your statement" (of the problem or fact), or "you couldn't have expressed the fact more correctly." If we want to call attention to our agreement with the *speaker,* then we are more likely to say, "you are correct as usual," "I couldn't agree with you more," "you hit the nail on the head," and so on. I do not deny that predicate adjectives have the secondary functions of indicating agreement with the speaker and/or his assertion, but I maintain this is not their primary function. Yet even though our agreement is with the stated fact, it is the statement that we call true or false because it bears the truth claim by virtue of depicting the fact or state of affairs.

Accordingly, this, I think, is what we ordinarily *mean* when we use terms such as 'true'; we intend to indicate agreement that the fact or state of affairs as designated by the statement is the case, and thus that the statement is true: that it "corresponds" to the facts. Moreover, I believe this intepretation is the one most consonant with the tremendous significance and value we attribute to the truth. "We thirst for the truth." "Above all, truth is to be cherished." "There is no compromising the truth." "We shall pursue the truth to the end." Why do we attach so much importance to the truth? Why are we incensed when someone deliberately misrepresents a situation or distorts the truth, outraged when someone has been mistakenly condemned? Is it merely because we are in disagreement with the person making the judgment or with the judgment made? No! Our outrage is based on the fact that the person has distorted reality, has lied in the face of the universe. Our response is not to a mere disagreement, but to a fundamental betrayal of man's instinctive reliance on the truth. To be sure, due to human weakness or out of consideration for others we do compromise our expression of the truth, but I do not think any theory of truth can be adequate that does not attest to man's deep yearning for the truth, however elusive its attainment may be!

This brings us to the question of how we determine when the truth conditions have been fulfilled: how we decide whether what has been asserted about the world is true or not? In traditional terms, what are the criteria and how do they function in assessing various truth claims? While the usual approach has been to single out a particular criterion of truth and defend it over any other, I shall attempt to justify the position that we are warranted in using different criteria in different circumstances. As Kant states:

> Now a general criterion of truth must be such as would be valid in each and every instance of knowledge, however their objects may vary. It is obvious, however, that such a criterion [being general] cannot take account of the [varying] content of knowledge....But since truth concerns just this very content, it is quite impossible, and indeed absurd, to ask for a general test of the truth of such content. A sufficient and at the same time general criterion of truth cannot possibly be given.[26]

I will try to demonstrate, therefore, focusing on the three major criteria of truth—correspondence, pragmatism, and coherence—that each has its own application within certain contexts or under certain conditions and limitations, so that there is no single unique or general criterion of truth. This view is mentioned by Sellars, but not developed by him.

> With the apparent achievement of a semantic *definition* of truth, 'pragmatic success', 'coherence', and (if there is a place for it) 'corre-

spondence'...no longer appear as mutually exclusive claimants to be the essence of truth. They might well, however, be compatible *properties* of truth or essential features of different varieties of truths.[27]

In addition, I will try to demonstrate that while each criterion has its proper application under the appropriate conditions, the other two are presupposed to some degree. Because it is closely related to the meaning of truth, I will conclude this chapter with a discussion of the correspondence criterion.

The Correspondence Criterion of Truth: According to the correspondence criterion (in its most sensible formulation), a statement is true if what it asserts "agrees with" or "conforms to" that which is asserted. This agreement, conformity, or correspondence is the test and mark of a statement's being true. But how is this test applied, and of what does its application consist?

It has long been recognized that language has two essential dimensions, or, as they have been called, "modes of meaning," which correspond somewhat to Austin's distinction between "descriptive" and "demonstrative conventions" (p. 22). One mode refers to the conventional meanings associated with individual words and their combined uses in grammatically correct syntactic units, and has been designated the "connotative" or "intensional meaning" (see Chapter II). These are the literal meanings we learn when we look up the definition of a word or define a foreign word in terms of our own language, and thus presuppose the meanings of other words. The second mode of meaning refers to the referential or designative use of language, and is called the "denotative" or "extensional meaning." This mode is prior, since it is by ostensive definition, by pointing to or designating things, that a child originally learns the meaning and application of its first words. For a language to be fully intelligible, therefore, we must understand both of these modes of meaning, as I argued in Chapters II, III, and IV.

This explains the frequent, if inexplicit, recourse in ordinary language to the correspondence criterion of truth: connotative or intensional meaning accounting for our initial understanding of what has been asserted, denotative or extensional meaning accounting for our ability to identify the states of affairs or features of the world designated by the assertion, whose actuality or non–actuality constitutes the confirmation or disconfirmation. As Paul Feyerabend states, describing a familiar situation is, for the speaker, an event in which statement and phenomenon are firmly glued together....From our very early days we learn to react to situations with the appropriate responses, linguistic or otherwise. The teaching procedures both *shape* the 'appearance', or 'phenomenon', and establish a firm *connection* with words, so that finally the phenomena seem to speak for themselves without outside help or extraneous knowledge. They *are* what the associated statements assert them to be.[28]

One could not be said to understand a statement fully if he could not identify its referent—if he could not identify, from his understanding of the statement, the state of affairs to which it refers. For example, if one says "it is snowing outside" or "you left the light on in the study," in each case, understanding the statement *implies* knowing the state of affairs it is about. Thus the correspondence criterion does not require taking a statement or an assertion and comparing it, like a photograph or a sample, with some objective reality; rather, it consists simply in *understanding* the statement and then *observing* whether what the statement depicts or designates occurs or not. In the above examples, one sees or feels the falling snow, or one observes a glow of light coming from the study. Since it is by means of language that we designate possible referents or states of affairs, there is usually no difficulty in ascertaining whether the two match up. As Strawson succinctly states: "Of course statements and facts fit. They were made for each other" (*op. cit.,* p. 39).

These examples illustrate the conditions under which the correspondence criterion is normally applied, indicating its usefulness as well as its limitations. Although its application is usually implicit, it is applied to ordinary statements within the context of a natural language with built-in criteria (contained in its different modes of meaning) for recognizing its referents. And since natural languages develop and are used under the conditions of our 'direct' experience of the world, no additional effort or experimentation is usually required for the confirmation of ordinary statements beyond that of simply looking, hearing, or ascertaining. In fact, the ordinary person would be puzzled or irritated if he were asked how he knew that statements such as "the sun is shining," "roses are fragrant," or "this coffee is cold" were true. He would probably answer that he just sees or experiences that they are true, and his reply would be correct—to doubt or deny the possible truth of such statements would be to question the conventions underlying the use of ordinary language.

It was the awareness of the seemingly transparent truth of such propositions, as well as the incontrovertible function of ordinary language in forming our common sense view of the world, that apparently led G. E. Moore to his classic "Defence Of Common Sense." In opposition to many scientists and philosophers who, for theoretical reasons, have denied the ultimate reality of such common features of the world as space, time, matter, and consciousness, Moore maintained that ordinary statements implying the reality of such features are so fundamental to our conception of existence that they could not be denied without self-refutation. As he says in his well-known article:

> . . . I am one of those philosophers who have held that the "Common Sense view of the world" is, in certain fundamental features [e.g., as regards space, time, matter, and conscious selves], *wholly* true. . . . The features in question. . .are all of them features, which have this peculiar property—namely, that *if we know that they are features in the "Common Sense view of the world," it follows that they are true*: it is self-contradictory to maintain that *we* know them to be features in the Common Sense view, and that yet they are not true; since to say that we know this, is to say that they are true.[29] (Brackets added.)

According to Moore, if the members of a linguistic community claim to *know* certain types of statements *to be true with certainty,* this is grounds for concluding that the *Weltanschauung* or world view implicit in those statements reflects necessary features of the real world. This follows not only from the fact that one "knows" such statements to be true, but also because if one denies their truth, and hence the features of the world implied by them, he necessarily contradicts himself. Examples of such statements are: "this is a hand," "the earth existed before I was born," "I am seated in front of my desk," and "I am thinking about what I am writing."

Moore claims that each of us knows such statements (and the class of statements they represent) to be true with certainty: that no one could 'really' doubt the existence of his hand, the prior existence of the earth, the fact that he exists in spatial relations to other objects, that occasionally he thinks, and so on. Because these statements imply the existence of certain features of the world, among them physical objects like hands and desks, temporal relations such as being born at a particular time, spatial relations such as being seated in front of something, and conscious experiences such as thinking, *these features must be real.* In addition, their reality is further proved by the fact that if one denies any of these classes of statements, he will inevitably contradict himself, since he will be denying that there is a world of objects (including his body), that he began to exist at a certain time, that he stands in spatial relations to other objects, or that he is conscious. If none of these facts were true, he could not exist to deny them.

While Moore's article does raise the Strawsonian question whether our thought about the world as expressed in ordinary language is so fundamental that it necessarily exhibits features of reality,[30] his argument that *because* such features are inherent in our common sense view, they *must* represent reality, is too facile. Kant, for example, did not deny that *for any human experience* phenomena must appear as physical objects in spatial and temporal relations, but he did deny that these features would be characteristic of reality *apart* from human experience. Similarly, scientists and philosophers of science (such as Adolf Grünbaum and Henry Mehlberg) who deny the asymmetry of time in the physical universe, do not deny "time's arrow" or the unidirectional flow of time as regards human experience. It is only if one confines himself to one context, such as that of ordinary experience, that this context will appear absolutely or unconditionally necessary.

As Whorf showed, as long as people knew only their own family of languages it was natural to assume that their language mirrored the world as it is; but when linguists compared the grammars of radically different languages, it became apparent that different languages projected different world views.[31] Analogously, before the development of the telescope and the microscope in the seventeenth century most people assumed that our eyes disclose the physical world as it is—one of the prejudices that stood in the way of accepting Galileo's telescopic observations.[32] The possibility of cross–cultural, cross–contextual comparisons reveals the limited nature of a particular context.

Moore used the paradigm case argument to the effect that fundamental linguistic conventions of the English language are necessarily true, and hence must reflect real features of the world. But Whorf found that the conceptions of space and time so basic to the Western world view did not occur in the same sense in the Hopi linguistic *Weltanshauung* at all. Accordingly, if Whorf's linguistic relativity thesis is true, what right has one to infer that the linguistic conventions of one's own language imply necessary features of reality, even when these conventions are so deeply embedded within a particular linguistic system as to appear absolutely necessary and inviolable to its users?

Moreover, one can offer examples of any number of statements which are transparently true *within a given context,* but which from a broader perspective or deeper context are seen to be true relative to the initial context only. To use a previous example, from our usual geocentric perspective such statements as "the earth is stationary," "the sun rises and sets," and "the stars appear after sunset and disappear at dawn," seem to be unconditionally true, but from a different and truer perspective, one knows they are true of appearances only (just as retrograde motions are not real motions of the planets but only apparent motions from the geocentric perspective.) Similarly, at the moment we are looking at the sun, it is natural to say "we are seeing the sun as it is now," when in fact we are seeing it after a lapse of eight minutes. Also, the statement "this drop of blood is red" has the same degree of certainty within the context of ordinary perception as Moore's favorite example, "this is a hand," yet we know that in a microscopic context the drop of blood does not appear red at all— again illustrating the importance of different conditions or contexts in disclosing the conditional nature of certain truths.

Developments in twentieth century physics, in particular, have pointed up the limited or conditional status of what had previously seemed to be incontrovertible common sense beliefs—illustrating the inevitable poverty of a philosophy committed to a defense of common sense and of ordinary linguistic uses. This is especially true of relativity theory which showed that basic concepts in Newtonian science that seemed to have an intuitive certainty and absolute status were true only within "limiting conditions," those in which the distances and velocities are negligible (as they are for all practical purposes on the earth) compared to the velocity of light. Before relativity theory, if someone had claimed that clocks would slow down, measuring rods contract, and mass increase in systems approaching the velocity of light, he would have been considered a "crackpot," yet today scientists accept such consequences as perfectly natural given the constant velocity of light.

It is only from the standpoint of different perspectives, conditions, or contexts that one becomes aware of the limitations of any particular context, whether that of Newtonian mechanics or of everyday experience. If all one knows is ordinary language (as was true of ordinary language philosophers such as Moore and Ryle), then naturally he is apt to accept the conventions of this language and the world view implied by it as absolute. Like those confined to the cave in Plato's allegory, Moore accepted the phenomena of immediate experience as being ultimately real. This is not to claim, however, that ordinary language does not have its appropriate use. On the contrary, I have been insisting that ordinary language does have a valid application *within the ordinary context of one's direct experience of the world,* so that given this context and the conventions of a natural language, the correspondence criterion of truth best represents the criterion we use in determining the truth of its statements (just as the prisoners in Plato's cave allegory could argue about, and in a limited way since they could not move about, visually confirm their descriptions of the reflected images).

Within the context of ordinary experience, therefore, it would be misleading to claim, as C. I. Lewis did, that common assertions such as "the paper on which I am writing is white," "the surface of my desk is hard," or "the shape of my pencil is hexagonal" require the application of the pragmatic criterion of truth.[33] To be sure, the truth of such statements transcends the initial evidence, but the claim that they are hypotheses whose consequences, when determined, will either confirm or disconfirm them, seems artificial, contravening the difference between actual hypotheses and ordinary assertions. We know ordinary statements to be true by observing the phenomena they designate, not by manipulating conditions to verify them, as required by the normal application of the pragmatic criterion of truth. Similarly, it would be factitious to claim that the truth of ordinary statements depends upon the coherence criterion, though such assertions are made on the basis of a particular conceptual–linguistic system, and therefore are confirmed *within* the context of that system.

However, such considerations do indicate the limitations of the application of the correspondence criterion, in that it applies especially to common assertions made in natural languages which are descriptive of ordinary phenomena and events as experienced under normal conditions, the conditions that prevailed when those particular linguistic conventions or modes of meaning were established. But while this criterion takes into account the necessity of a background language or cognitive–linguistic system for making any truth claims, it uncritically assumes that the conjunction of the two in a *particular* natural language

(Moore's mistake) is a true representation of reality—that a particular universe of discourse necessarily mirrors reality.

However, as any consideration of more primitive world views illustrates, there is no guarantee that we may not be systematically misdescribing or misconceiving the world at any one time, regardless of how self-evident or incontrovertible the evidence and rationale may appear. Intellectual history is a graveyard of *Weltanschauungs* that were one time as vital "live options" as our currently accepted views. Even the much touted self-corrective process inherent in the methodology of science is effective relative to that methodology, just as during the Middle Ages divine revelation and faith were accepted as the "highest forms of knowledge," on the presumed authority of God. Nevertheless, we have no recourse other than to rely on what seems to be the best cognitive-linguistic system available at any time in communicating our judgments and beliefs about the world. Within the context of that system truth claims can be adjudicated, even though what is considered true relative to that framework may later be falsified, or even considered false at the time by those who reject that framework (such as the creationist's rejection of the theory of evolution). Paradoxical as this may sound, it is a natural consequence of the growth of knowledge. As we learn more about the world, both in terms of new discoveries and the refinement or revision of concepts and their implications, the referents and states of affairs that confirm knowledge also undergo a change. And though the most revolutionary of these changes may seem incommensurable at the peak of the controversy, they invariably appear less abrupt and disparate from a longer historical perspective, which explains why the development of science resembles more a gradual approximation of a deeper and more comprehensive understanding of physical reality, than a series of alternative, disconnected episodes.

Furthermore, because the correspondence criterion is used more frequently in ordinary discourse, where the correspondence between the assertions and what they designate is usually unproblematic, given their conventionally accepted modes of meaning, it is less applicable, or even inapplicable, in situations where the confirming states of affairs are unknown or even non-existent at the time a truth claim is made, or which can never be directly experienced (as in particle physics). It is in the latter contexts, as we shall find, that we rely on either the pragmatic or the coherence criterion as the test of truth.

Notes to Chapter VI

1. Richard Rorty, *Philosophy and the Mirror of Nature* (Princeton: Princeton University Press, 1979), p. 161.
2. Wilfrid Sellars, *Science, Perception and Reality* (New York: Humanities Press, 1963), p. 169.
3. Cf. John R. Searle, *Intentionality* (Cambridge: Cambridge University Press, 1983), pp. 18–19.
4. Cf. R. Edgley, "Chomsky's Theory of Innate Ideas," in *Challenges to Empiricism,* ed. by Harold Morick (Indianapolis: Hackett Publishing Co., 1980), p. 273.
5. Aristotle, *Metaphysics,* 1011b:25–28. Trans. by Ross.
6. Cf. Paul Churchland, *Scientific Realism and the Plasticity of Mind* (Cambridge: Cambridge University Press, 1979), p. 25.
7. Cf. Karl R. Popper, *Objective Knowledge* (Oxford: Clarendon Press, 1972), p. 71.
8. Cf. W. V. Quine, *Ontological Relativity and Other Essays* (New York: Columbia University Press, 1969), pp. 48–49.
9. Cf. Wilfrid Sellars, *op. cit.,* p. 160.
10. Cf. Paul Feyerabend, *Against Method* (London: Verso Edition, 1975), p. 72.
11. Cf. Immanuel Kant, *Prolegomena to Any Future Metaphysics,* revised Carus trans. (Indianapolis: Bobbs–Merrill, 1950), pp. 99–113.
12. Cf. Immanuel Kant, *Critique of Pure Reason,* trans. by Norman Kemp Smith (London: Macmillan & Co., 1929), p. 271.
13. Cf. Benjamin Lee Whorf, *Language, Thought, and Reality,* ed. by John B. Carroll (New York: MIT Press and John Wiley & Sons, 1956), pp. 57–64, 207–219.
14. Cf. Karl R. Popper, *op. cit.,* ch. 1.
15. Cf. Albert Einstein, "On the Electrodynamics of Moving Bodies," in *The Principle of Relativity,* H. A. Lorentz, A. Einstein, H. Minkowski, and H. Weye, trans. by W. Perrett and G. B. Jeffery (New York: Dover Publications, 1923), pp. 37–65.
16. Cf. Roy Bhaskar, *A Realist Theory of Science,* 2nd ed. (Sussex: The Harvester Press, 1978), pp. 249–250. He presents a similar position, but denies the correspondence criterion of truth.
17. Cf. Bertrand Russell, "Descriptions," and P. F. Strawson, "On Referring," reprinted in *Classics in Analytic Philosophy,* ed. by Robert Ammerman (New York: McGraw–Hill, 1965), ch. 1, ch. 14.

18. Cf. Alfred Tarski, "The Semantic Conception Of Truth," *Philosophy and Phenomenological Research,* 4, 1944. Reprinted in Leonard Linsky, ed., *Semantics And The Philosophy Of Language* (Urbana: University of Illinois Press, 1952), pp. 13–47. While the main point of Tarski's article was to avoid semantic paradoxes such as the "antinomy of the liar," and therefore was chiefly concerned with semantics, I shall focus on the problem of truth itself. Tarski's article is so well known and has been so thoroughly discussed that it hardly needs repeating. In addition to Popper who dedicated *Objective Knowledge* to Tarski, Sellars devotes a chapter to his conception of truth in *Science, Perception and Reality.* For more recent discussions, see Hilary Putnam's *Mind, Language and Reality,* Vol. 2 (Cambridge: Cambridge University Press, 1975), ch. 3; Mark de Breton Platts, *Ways of Meaning* (London: Routledge & Kegan Paul, 1976), chs. 1, 2; David Papineau, *Theory and Meaning* (Oxford: Clarendon Press, 1979), ch. 3; and Saul Kripke, "Outline of a Theory of Truth," *The Journal of Philosophy,* Vol. LXXII, No. 19, November 6, 1975, pp. 690–716.

19. Leonard Linsky, (ed.), *Semantics And The Philosophy of Language, op. cit.,* p. 15.

20. Max Black, *Language and Philosophy* (Ithaca: Cornell University Press, 1949), p. 91. Brackets added.

21. Cf. *ibid.,* p. 104.

22. Cf. "Truth," *Proceedings of the Aristotelian Society,* Supp. Vol. XXIV, 1950.

23. Cf. John Austin, "Truth," in *Truth,* ed. by George Pitcher (Englewood Cliffs, N.J.: Prentice Hall, 1964), p. 19. Subsequent references to this article will be indicated in the text by page number.

24. P. F. Strawson, "Truth," in *Truth, op. cit.,* pp. 52–53. Subsequent references to this article will be indicated in the text by the page number.

25. G. J. Warnock, "A Problem About Truth," *Truth, op. cit.,* p. 59. Italics added.

26. Immanuel Kant, *Critique of Pure Reason, op. cit.,* pp. 97–98.

27. Wilfrid Sellars, *op. cit.,* p. 198.

28. Paul Feyerabend, *op. cit.,* p. 72.

29. G. E. Moore, "A Defense of Common Sense," in *Contemporary British Philosophy,* second series, ed. by J. H. Muirhead (London: George Allen & Unwin, 1915). Reprinted in Robert Ammerman, *op. cit.,* quoted from p. 56.

30. Cf. P. F. Strawson, *Individuals: An Essay in Descriptive Metaphysics* (London: Methuen & Co., 1959).

31. Cf. Benjamin Lee Whorf, *op. cit.,* pp. 57–58.

32. Cf. Ludovico Geymonat, *Galileo Galilei,* trans. by Stillman Drake (New York: McGraw-Hill, 1965), p. 45ff.

33. Cf. C. I. Lewis, *Mind And The World Order* (New York: Dover Publications, 1929), p. 50.

CHAPTER VII

THE PRAGMATIC CRITERION OF TRUTH

AS the discussion of the correspondence criterion in the previous chapter indicated, the use of a particular test of truth presupposes certain experiential and linguistic conditions constituting a particular world view. Assuming the world is directly accessible, and that the conceptual–linguistic framework in use mirrors the world as it is, one normally relies on the correspondence criterion, "so that he who says of anything that it is, or that it is not, will either say what is true or what is false. . . ." As Aristotle's statement implies, these assumptions were integral to the preeminent Classical Greek tradition, consisting of knowledge of the Real Essences or Substantial Forms inherent in things, on his view, or of the Eternal Forms subsisting in a Transcendent Realm, on Plato's. Moreover, the belief in an independent, self-subsistent universe, knowable at least in principle, provided the ontological setting for most scientific inquiry and philosophic speculation in the West, until recently, as Dewey persuasively argued in the *Quest For Certainty*.[1]

Though its full implications were not apparent until the present century, it was the emergence of modern science that led to the destruction of the classical world view and its replacement by a different conception of physical reality. According to the latter, the

actual world is not revealed in the qualitative Essences or manifest
Forms of observable phenomena (which are considered mind-
dependent on the new scheme of things), but consists of the motions and arrangements of the "insensible particles" or atoms underlying natural phenomena. Thus the ancient Greek tradition of
atomism developed by Leucippus and Democritus,[2] but eclipsed by
the philosophy of Plato and the cosmology of Aristotle, reemerged
as the dominant world picture. As Newton declares in a famous
passage:

> All these things being consider'd, it seems probable to me, that God
> in the Beginning form'd Matter in solid, massy, hard, impenetrable,
> moveable Particles, of such Sizes and Figures, and with such other
> Properties, and in such Proportion to Space, as most conduced to the
> End for which he form'd them; and that these primitive Particles
> being Solids, are incomparably harder than any porous Bodies com-
> pounded of them; even so very hard, as never to wear or break in
> pieces; no ordinary Power being able to divide what God himself
> made one in the first Creation.[3]

All the composition and changes of nature were attributed to the
motions and forces of these insensible atoms or corpuscles.

Though the real world of atoms and forces lying behind the world
of appearances was inaccessible to direct observation, it occupied
Newton's absolute "frames of the universe," space and time. Space
was defined as an infinite, Euclidean, isomorphic container logically
presupposed by the idealized dimensionless particles (assumed for
computational purposes) or point masses scattered throughout the
universe. Owing to the absolute structure of space, these point
masses, physically interpreted as indivisible atoms, had a fixed
location if at rest and an absolute displacement if in motion. Similarly,
time (though measured by terrestrial creatures in conventional units
based on the earth's diurnal rotation and the orbital periods of the
moon and the sun), in itself flowed invariantly and eternally, independent of any physical motions. Although Newton seemed to be
aware of the possibility of Einsteinian variations in temporal durations, he denied them: "All motions may be accelerated and retarded,
but the flowing of absolute time is not liable to any change. The
duration or perseverance of the existence of things remains the same,
whether the motions are swift or slow, or none at all. . . ."[4]

The indestructible atoms composing this world machine were
defined by the "primary qualities" of position, extension, size,

shape, motion or rest, and solidity or mass, conceived as intrinsic, invariable properties. Forces such as gravity, inertia, impact, and momentum, though barely understood, were considered sufficient to provide (in principle) a deterministic explanation of all motion and change. It was not the laws of nature but the inaccessibility of the atoms, the vagueness of the concepts of forces, and the obscurity of the mechanical processes that gave rise to the prevalent skepticism of the eighteenth century, epitomized by Hume, described in Chapter IV. Nonetheless, it was the remarkable discoveries and progress based on the Newtonian paradigm in the ensuing two centuries, such as the theoretical prediction of the planet Neptune to explain the irregular orbit of Uranus, subsequently confirmed, that led, by the end of the nineteenth century, to such confidence in its ultimate and universal validity.

This belief in the Newtonian mechanistic world picture and confidence in the absoluteness of deterministic explanations was shattered by developments in relativity theory and quantum mechanics. Although relativity theory did not deny causal determinism, it did refute the independence and invariance of Newton's absolute frames of the universe, and showed that the seemingly intrinsic, invariant metrics (within terrestial dimensions) of extension, duration, and mass were variable as the system approaches the velocity of light and/or as measured from coordinate systems in relative motion.[5] But it was quantum mechanics, based on discontinuous, discrete quanta of radiational interactions and the discovery of indeterminate conjugate properties, probabilistic processes, irreducible dualities, and inscrutable interactions at the subatomic level, that drastically undermined the absolute validity of Newtonian mechanics. As described by David Bohm in his classic book on quantum mechanics:

> First, the classical concept of a continuous and precisely defined trajectory is fundamentally altered by the introduction of a description of motion in terms of a series of indivisible transitions. Second, the rigid determinism of classical theory is replaced by the concept of causality as an approximate and statistical trend. Third, the classical assumption that elementary particles have an "intrinsic" nature which can never change is replaced by the assumption that they can act either like waves or like particles, depending on how they are treated by the surrounding environment. The application of these three new concepts results in the breakdown of an assumption which lies behind much of our customary language and way of thinking; namely, that the world can correctly be analyzed into distinct parts,

each having a separate existence, but working together according to
exact causal laws to form the whole.[6]

These consequences are so radical in terms of previous assumptions and conceptions that all kinds of ingenious or fabulous interpretations have been suggested to account for them: that our
traditional two-valued logic must be replaced by a multi-valued
"quantum logic;"[7] that the nature of subatomic processes are such
that their source must lie outside of space-time and independent of
the principle of causality;[8] that the effects of the investigation on
one part of a system are conveyed faster than the speed of light (in
violation of relativity theory and local causality) to another, so that
the second part seems to know "telepathically" or instantaneously
what is happening to the first;[9] that there are "other worlds" (the
Everett-Wheeler-Graham theory) within which the alternative or
"branch processes" not actualized in this world are realized;[10] or
that the anomalies of subatomic physics indicate the ultimate limitations of experimental investigations and of rational explanation
(that is, of western science), along with the illusion of all macroscopic separateness and independence, implying that ultimate
reality must be grasped as a transcendent unity, similar to the claims
of eastern mysticism and western absolute idealism.[11]

While the dismantling of Newton's mechanistic universe along
with the relativization of physical properties and the introduction
of indeterminism contributed to the development of pragmatism
(as indicated in the expositions of Dewey and Lewis), the more
basic source was the appreciation of the significance of the experimental method of science and its practical consequences. If the
states, properties, and causal interactions of phenomena are not
immutable and self-subsistent, but depend instead upon certain
conditions and relations that can be discovered experimentally,
then this allows for some modification and control of the phenomena. A universe so constituted would explain the necessity for,
and possibility of, experimental inquiry, along with the import of its
consequences. For if we understand what causes natural events and
why objects have the properties and effects they do, this understanding enables us to redirect occurrences and modify the effects
of things in accordance with our needs and desires. No longer are
we "fated" to accept nature and events as completely "destined." As
Dewey states:

> The central and outstanding fact is that the change in the method of knowing, due to the scientific revolution begun in the sixteenth and seventeenth centuries, has been accompanied by a revolution in the attitude of man toward natural occurrences and their interactions. This transformation means...a complete reversal in the traditional relationship of knowledge and action. Science advances by adopting the instruments and doings of directed practice, and the knowledge thus gained becomes a means of the development of arts which bring nature still further into actual and potential service of human purposes and valuations (p. 85).

It was because of their knowledge and appreciation of science that Peirce, James, and Dewey used science as their model of knowing, introducing and defending the new position of pragmatism. Peirce, the founder of pragmatism, was an accomplished mathematician, astronomer, and research scientist for the Geological Society; James was trained as a physician and psychologist; and Dewey was much better informed of developments in the sciences, from anthropology and biology to physics, than were most of the philosophers of his day.

According to Dewey's version of pragmatism, thought or reflection begins when one is confronted by a "problematic," "unsettled," or "indeterminate" situation. The classical empiricist view of ideas as representations of an independent, antecedent reality was a mistake, since ideas function as hypotheses or tenatative plans of action proposed as possible solutions to the problematic situations. They are "operational" in directing actions or experimental procedures intended to resolve the initial doubt, and in so far as they lead to the resolution of the problem, they are "instrumental." As inquiry proceeds, the original ideas or hypotheses are progressively redefined to meet the experimental requirements until the original problem is solved, indicated by the elimination of the initial perplexity. According to the pragmatic criterion, a statement expressed as a hypothesis intended to resolve a troubling situation is true or false, depending upon whether the consequences that ensue from acting on the hypothesis do or do not solve the problem. As Dewey states, "We know...whenever our inquiry leads to conclusions which settle the problem out of which it grew" (p. 198).

On this view the mind is not, as in classical theories, a faculty for disclosing an immutable reality, but an organ for facilitating changes in the environment to effect more successful adaptations to the world, and therefore is oriented toward future consequences, rather than past actualities. The test of truth does not consist in the matching of statements with conventionally established facts or states of affairs, but in whether the consequences resulting from implementing the statement eventuates in successfully altering a problematic situation so as to eliminate the difficulty. Accordingly, we utilize the pragmatic criterion when the verifying state of affairs, rather than being previously known and preexistent, is a conjectured or indeterminate state *the actualization of which* depends upon acting on various hyopotheses. This functional view of thought as submitting proposed hypotheses to competitive tests or selection, in our ongoing attempts to understand and control the environment, is seen as a continuation of the evolutionary process.

But while pragmatists like Dewey and Lewis emphasized the conditional or relational status of natural phenomena, and the consequent change in the conception of knowing brought about by the experimental method in science, they were overly committed to common sense realism and instrumentalism. Dewey was so opposed to the traditional problems of epistemology and any form of dualism that he was unwilling to acknowledge that the discoveries of science disclose additional dimensions of physical reality in some respects more basic than the world of ordinary experience. As he says, "the claim of physical objects, the objects in which the physical sciences terminate, to constitute the real nature of the world, places the objects of value with which our affections and choices are concerned at an invidious disadvantage" (pp. 195–196). Thus he was adverse to recognizing a deeper domain of physical structures beyond that of the qualitative macroscopic world, and settled for an instrumentalist interpretation of truth, as "warranted assertibility." Like his contemporary successors such as Richard Rorty, Larry Laudan, and Arthur Fine, Dewey neglected to press far enough the question why experimental inquiry was so successful in resolving problems. As he typically says:

> The search for "efficient causes" instead of for final causes, for extrinsic relations instead of intrinsic forms, constitutes the aim of science. But the search does not signify a quest for reality in contrast with experience of the unreal and phenomenal. It signifies a search for those relations upon which the *occurrence* of real qualities and values depends, by means of which we can regulate their occurrence (pp. 103–104).

For Dewey, as for all pragmatists, the aim of science is to discover the "relations" or "conditions" on which the ordinary objects of experience possessing qualities and potential values depend, so "as to transform distinctly human values in behalf of a human interest" (p. 199). Scientific knowledge is never an end in itself as a quest for further understanding the universe, but is merely instrumental in providing greater means of control over the experienced world. Again, the reason he denies ontological status to the entities discovered by science, except as "relations," is that he does not want them to be taken as more real than the objects of ordinary experience.

> The relations a thing sustains are hardly a competitor to the thing itself. Put positively, the physical object, as scientifically defined, is not a duplicated real object, but is a statement, as numerically definite as is possible, of the relations between sets of changes the qualitative object sustains with changes in other things—ideally of all things with which interaction might under any circumstances take place (pp. 130–131).

We should thus "surrender the traditional notion that knowledge is possession of the inner nature of things and is the only way in which they may be experienced as they 'really' are" (p. 131).

In his concern to preserve the preeminence of the ordinary world, relegating the experimental discoveries of science to relations sustained by qualitative objects undergoing change, Dewey's position is the opposite of that of Paul Churchland. According to Dewey, "[i]t seems almost too obvious for mention that a scientific object consisting of a set of measurements of relations between two qualitative objects. . .cannot possibly be taken, or even mis–taken, for a new kind of 'real' object which is a rival to the 'reality' of the ordinary object" (p. 130). "Water as an object of science, as H_2O with all the other scientific propositions which can be made about it, is not a rival for position in real being with the water we see and use" (p. 106). In contrast, Churchland maintains:

> The class of perceptual beliefs must now be counted as a subclass of the class of theoretical beliefs: roughly, as those singular theoretical beliefs acquired as spontaneous non–inferential responses to sensory states of the perceiver. As with singular theoretical judgements generally then, the adequacy of our perceptual judgements is in part a matter of the adequacy of the background theory (conceptual frame-

work) in whose terms they happen to be framed. Our perceptual judgements can no longer be assigned any privileged status as independent and theory–neutral arbiters of what there is in the world. Excellence of theory emerges as the fundamental measure of all ontology. The function of science, therefore, is to provide us with a superior and (in the long run) perhaps profoundly different conception of the world, *even at the perceptual level.*[12]

While I think, as I argued in Chapter I, that Churchland's claim that as science progresses our conception of ourselves as sentient beings, and of the perceptual world as qualitative objects, will gradually be replaced by purely scientific concepts is unlikely, Dewey's notion of scientific discoveries as not challenging, in any way, the status of ordinary objects also seems to be naive.

I believe there are two major reasons why pragmatists such as Dewey and Lewis adopted antirealist interpretations of science and instrumentalist theories of truth (although Lewis' "conceptual pragmatism" was qualified by a Platonic notion of *a priori* truths and of the importance of coherence as a measure of truth[13]). One reason was the pervasive influence at the time of a positivistic interpretation of science. James' Humean concept of "pure experience" was very similar to that of Mach, the influential Viennese physicist whose book, *Die Mechanik in Ihrer Entwicklung Historisch-Kritisch Dargestellt* (translated *The Science of Mechanics: A Critical and Historical Account of Its Development*), went through nine German editions from 1883 to 1933, and had an enormous influence. James met Mach in Vienna and praised him to the skies. The fact that the epistemological tenets of the positivists could be interpreted as being similar to those of the founders of quantum mechanics, and that Einstein had generously acknowledged his *early* indebtedness to the positivistic doctrines of Mach and Poincaré, contributed greatly to the prestige and dominance of positivism among philosophers and scientists from the thirties through the sixties. Dewey himself contributed to the positivist's publication, *International Encyclopedia of the Unified Sciences.* Although we are now in the "post-positivist era," the legacy of positivism still exerts a subtle influence, as can be seen in Quine's notion of observation sentences, in the persistent acceptance of Hume's analysis of causality, and in the continued opposition to a realistic interpretation of science.

The second reason Dewey and Lewis adopted the interpretation of science they did was that many areas of science at that time were still in a relatively rudimentary stage of development. Although the special and general theories of relativity had been formulated and the classical phase of the development of quantum mechanics had been achieved by the time Dewey had written *The Quest for Certainty* and Lewis *Mind And The World Order,* both published in 1929, the major experimental developments in science were yet to come: the production of "wonder drugs" such as penicillin and vaccines for preventing and/or curing diseases; discoveries in neurophysiology providing greater knowledge of the anatomical structures of the brain and their functions, along with an understanding of the chemical-electrical nature of neuronal and synaptic transmissions, with the imminent likelihood of curing such devastating diseases as Parkinson's, Alzheimer's, and dementia praecox that afflict the elderly; developments in pharmacology alleviating physical disorders (such as epileptic seizures) and diseases, with the possibility of controlling such debilitating mental conditions as manic-depression and schizophrenia; the discovery of the molecular structure of the gene creating the fields of molecular biology and genetic engineering which produced the "green revolution" and the means of eliminating horrible genetic defects such as the Lesch-Nylan syndrome; developments in physical and organic chemistry providing greater knowledge of the chemical structure of inorganic substances and organic systems, and leading to the production of plastics, synthetic fibers such as nylon, and artificial organs; the creation of atomic fission and fusion and the construction of atomic and hydrogen bombs, nuclear reactors and submarines; the building of CAT and PET scanners for diagnosing physiological abnormalities and the use of radiation and chemotherapy in the treatment of cancer; the development of computers, microchips, robots, and artificial intelligence facilitating such engineering feats as the lunar landing and exploration by satellite of Mars, Jupiter, and Saturn.

When I read Dewey, I am struck by the contrast between his astute and farsighted assessment of the effects of scientific discoveries on our traditional beliefs and ways of doing things, and his simplistic conception of the nature and results of experimental inquiry. All he ever refers to are the "relations," "conditions," and "consequences" uncovered by experiments, not what they specifically consist of. In a typical passage, he says "experimental inquiry or thinking signifies *directed activity,* doing something which varies the conditions under which objects are observed and directly had and by instituting new arrangements among them" (p. 123). This is true as far as it goes, but it does not go very far in describing the actual discoveries of experimental inquiry, how such discoveries occurred, and why they had such important practical consequences. The general answer to this, as the history of modern science indicates, is that while the world as we initially encounter it displays certain regularities, it is not self-explanatory. The underlying causes of observable phenomena are not disclosed to our senses which are restricted to limited ranges of stimuli, and therefore these unobservable causes, consisting of elements, structures, and forces, must be inferred and/or uncovered by experimentation. The history of modern science is a record of such inferences and discoveries.

The earliest dramatic instance of this occurred in the first half of the seventeenth century when Galileo, by his use of the telescope, observed for the first time the mountainous surface of the moon, the four (Medician) satellites of Jupiter, and the phases of Venus, the new evidence crucial for the support of Copernicus' heliocentric hypothesis. During the same period, Boyle and Hooke were using the air–pump to investigate the elasticity and compressibility of the air, while Torricelli constructed a mercury barometer to demonstrate that the atmosphere, although imperceptible, has weight. The development of the microscope in the second half of the seventeenth century considerably enhanced biological and physiological research by disclosing previously inaccessible structures, such as the "capillary vessels" connecting the arteries and veins posited by Harvey to explain the circulation of the blood. In addition, Leeuwenhoek and Malphighi used the microscope to observe the corpuscles of the blood and the fibrillary structure of muscles.

Extensive experiments were conducted in the eighteenth century by Black, Cavendish, Priestly, and Scheele culminating in Lavoisier's identification of oxygen to explain combustion and respiration. In the nineteenth century, the discovery by Proust of the constant composition of chemical compounds, and the investigation by Charles, Dalton, Gay-Lussac, and Avogadro of the reaction of gases to changes in pressure, temperature, and volume, and of the proportions by which they combine, culminated in the various gas laws and in the molecular interpretation of the combination of gases. The long tradition of optical investigations by Descartes, Kepler, Newton, and Huygens resulted in Snell's Law describing the refraction of light in denser mediums and Young's experiments demonstrating the interference or diffraction of light, confirming its wave character. Fresnel gave the undulatory theory of light mathematical precision by deriving quantitative measurements of wave lengths and amplitudes. Fraunhofer developed the spectroscope to examine the solar spectrum (with the now familiar transverse lines characteristic of specific colors or wave lengths), while Kirchhoff succeeded in demonstrating that an invariable correlation existed between spectral lines and certain kinds of matter.

Experimental investigations to explain the nature of heat and electricity were pursued throughout the eighteenth and nineteenth centuries. Franklin succeeded in demonstrating that lightning is of electrical origin and the discoveries of experimentalists like Faraday eventuated in Maxwell's equations of electromagnetism, one of the greatest achievements of the nineteenth century. Galvani accidentally discovered that nervous and muscular contractions could be induced by electrical stimulation. J. P. Joule and William Thomson (Lord Kelvin) defined the new principle of "energy," while the laws of classical thermodynamics were the work of Carnot, Mayer, Helmholtz, Thomson, and Clausius. Among the founders of statistical mechanics were Maxwell, Clausius, Gibbs, and Boltzmann. J. J. Thomson's investigation of cathode rays led to the identification and determination of the properties of the electron. Röntgen accidentally discovered X-rays, and the experiments of Henri Becquerel and the Curies demonstrated the existence of radiation and of radioactive substances, such as uranium. Rutherford's atomic experiments disclosed the existence of the subatomic nuclear particles, the proton and neutron, suggesting the "Saturnian model" of the atom, and initiating the type of research in particle physics that has had such dramatic consequences in the second half of the twentieth century.

In biology, naturalists such as Hunter, Linnaeus, and Cuvier were occupied with taxonomy, the classification of organism according to their structures. In the early part of the nineteenth century Purkinje introduced the concept of protoplasm as the substance of which cells are composed, Schwann was mainly responsible for the theory of the cell, and Virchow for the notion that tissues composed of cells make up the body. In 1859 (the year of Dewey's birth) Darwin published his great work, *The Origin of Species,* radically transforming man's conception of himself and making possible for the first time a completely naturalistic explanation of his origin. In the latter part of the century the great work in microbiology was due to Pasteur, while the improved techniques of investigation involving stains, microtomes, and oil immersion lenses led to a much more exact understanding of cell division, such as mitosis, and the mechanism of cellular reproduction. Weismann and Mendel contributed to the rise of the science of genetics.[14]

Although only a brief sketch, these scientific investigations, discoveries, and theories that occurred by the time of Dewey's Gifford Lectures in 1929, show how inexact was his characterization of scientific inquiry. Instead of vague references to "relations," "conditions," "actions," and the "instrumental" success of scientific theories, the crucial question is whether the experimental inquiries of science disclose the existence of *entities* beyond those of the objects of ordinary experience (such as Neptune, electrons and protons, cells and cellular structures, X-rays and radioactive materials), structures (such as the linear alignment of atoms in crystals and metals, the undulatory form of light, the molecular and biomolecular structures of HCl and DNA-RNA), *processes* (such as photosynthesis, chemical reactions, radiation, cellular replication), and causes (of scurvy, the bubonic plague, malaria, muscular contraction, and cancer). To laud the success of science and at the same time refuse to attribute any status to the entities, mechanisms, and processes *in terms of which this success was achieved and explained,* is inconsistent. It is, in fact, to attribute magical powers to science in the sense of being entranced by its performance but totally incapable of comprehending how or why the wonderous feats took place.

While Dewey can be somewhat excused from his loose understanding of the actual accomplishments of science because of the uncertainty regarding its achievements in his day, this is not true of such contemporary antirealists and instrumentalists as Arthur Fine and Larry Laudan. As for Fine, his emotional railings against scientific realism show little understanding of the history and methodology of science, a misinterpretation of the controversies in quantum mechanics, along with internal inconsistencies or ambiguities. In addition, Fine is prone to drawing false analogies. In the symposium on "Realism and Anti-Realism" at Bryn Mawr College in the fall of 1984, he maintained that the current justification for belief in scientific realism is analogous to the eighteenth century justification for the belief in God. However, as the latter was based primarily on the arguments from analogy and design, while the belief in scientific realism is supported by the success of the experimental and hypothetical-deductive methods in science, which are entirely different, Fine's argument betrays a certain cavalier treatment (or misunderstanding) of the issues involved. Yet in his article, "The Natural Ontological Attitude,"[15] he persists in equating belief in scientific realism with a religious faith. But the belief in scientific realism is no more a religious belief than the belief in antirealism, and even less so because the latter claims that we should accept the results of science on faith, not being able to account for their occurrence.

Fine draws a second erroneous analogy between the mathematical discussions of set theory and "the significance of the explanatory apparatus in scientific investigations" (p. 85). However, since mathematics does not include an experimental method for making empirical discoveries nor does it make any claims regarding the actual properties, states, or processes in nature, any conclusion based on an analogy between the two is *prima facie* suspect.

But much more serious is Fine's misconstrual of the basic issue between a realistic and an antirealistic interpretation of scientific inquiry. Instead of interpreting scientific realism as being *implied by* the achievements of science, Fine describes the position as presenting *an additional justification* of the success of science. As he says, "no support accrues to realism by showing that realism is a good hypothesis for explaining scientific practice" (p. 86). The "realist's failure...resides...in his repeating the question–begging move from explanatory efficacy to the truth of the explanatory hypothesis" (p. 89), and to the existence of its components or referents.

> We have...seen how these ideas of correspondence and approximate truth are supposed to explain what *makes* the truth *true* whereas, in fact, they function as mere trappings, that is, as superficial decorations that may well attract our attention but do not compel rational belief...they are an arresting footthump and, logically speaking, of no more force (p. 97).

Even if other philosophers defend scientific realism on the grounds of its being able to explain or justify the acknowledged successes and advances in scientific inquiry, that is not my position. Instead, I believe that the correct defense is to show that *accepting* the "successes" of science, while at the same time *denying* the entities and mechanisms that constitute this success, is contradictory. For example, is it consistent to accept the development of the atomic bomb and then deny the truth of the atomic theory and the existence of atoms and atomic fission? How can one explain the success of landing men on the moon with all the intricate calculations of thrust, acceleration, gravitational force, and orbital trajectories that this required, and yet maintain that our understanding of the laws of nature is unreliable? Yet Fine asks us to "[s]uppose...that the usual explanation–inferring devices in scientific practice do not lead to principles that are reliably true (or nearly so), nor to entities whose existence (or near–existence) is reliable" (p. 85). While he offers this as a supposition, how could the achievements of science have occurred on the basis of non-reliable principles and non-existent entities? As I indicated earlier, this requires a belief in the magical powers of science.

The only reason I can find for Fine maintaining the position he does is that he never asks the two obviously crucial questions: (1) how does one account for the success of scientific theories?, and (2) how could scientific theories be successful if they were false (as opposed to being provisional, imprecise, and approximate)? I would like Fine (or anyone else) to account for the explanatory success of theories while denying their *modus operandi*: Newton's universal law of gravitation *without* gravity; the theory of combustion and oxidation *without* oxygen; the corpuscular-kinetic theory of gases *without* corpuscles; molecular structures and reactions *without* molecules and chemical bonds; atomic fission, fusion, and radiation *without* atoms; photosynthesis *without* radiant energy and carbohydrates; the formation of chlorophyll *without* $C_{55}H_{72}MgN_4O_5$ and $C_{55}H_{70}MgN_4O_6$; neuronal discharges *without* charged ions of sodium (Na^+) and potassium (K^+); genetic engineering *without* DNA and RNA, and so on. In the case of Newton's law of gravitation, one could argue that the law has been successful even though there are no gravitational "forces" as Newton conceived them, but there still is something that accounts for the correlation of the orbital periods of the planets and their distances from the sun as described in Kepler's third law, whether it be the diminishing intensity of Newton's forces or Einstein's space-time field.

As Fine looks at the history of science it consists mainly of failures:

> Overwhelmingly, the results of the conscientious pursuit of scientific inquiry are failures: failed theories, failed hypotheses, failed conjec-

tures, inaccurate measurements, incorrect estimations of parameters, fallacious causal inferences, and so forth. If explanations are appropriate here, then what requires explaining is why the very same methods produce an overwhelming background of failures and, occasionally, also a pattern of successes. The realist literature has not yet begun to address this question, much less to offer even a hint of how to answer it (p. 104, f.n. 8).

One reason why the history of science, for Fine, appears to be a history of failures is that he accepts the Kuhnian-Feyerabendian thesis that there is no progression in science, only a succession of incommensurable theories replacing one another. He "is not committed to the progressivism that seems inherent in realism" (p. 98), apparently because he does not see "how to account for the successes of the later theories in new ground or with respect to novel predictions, or in overcoming the anomalies of the earlier theories" (p. 88). But which should the reader find more persuasive, the fact that the history of science clearly *does* illustrate these latter developments, or that Fine does not know "how to account for" them? This problem was addressed in Chapter IV and the answer will be underscored later in this chapter, so I will not pursue it further now. Suffice it to say that the continued successes of such powerful theories as the atomic, evolutionary, and neuronal belies the pessimism of Fine.

Like many other antirealists (Laudan is an exception), Fine bases his position mainly on developments in quantum mechanics—ignoring the unquestioned achievements of science in so many other fields—where the problem of the underlying reality implied by the experimental conditions and the mathematical formalism does raise serious problems of interpretation. But even his assessment of the EPR (Einstein, Podolsky, and Rosen) controversy with Bohr is largely misinterpreted. As Fine describes it, the controversy was a "war between Einstein, the realist, and Bohr, the nonrealist" (p. 93). He describes Bohr as "the archenemy of realism" (p. 96), and his "so-called 'philosophy of complementarity'" a "nonrealist position" (p. 93). But this construal is a self-serving distortion of the basic issues.

As the discussion in the next chapter of the famous EPR controversy between Einstein and Bohr will more fully illustrate, the disagreement was not over a realist and nonrealist interpretation of the experimental conditions and underlying reality in quantum mechanics, but whether the *classical picture* of physical reality should be maintained and quantum mechanics declared "incomplete," or whether one should give up the classical picture and reinterpret physical reality to conform to the experimental conditions and mathematical formalism of quantum mechanics. This is explicitly stated by Bohr:

> Indeed the *finite interaction between object and measuring agencies* conditioned by the very existence of the quantum of action entails—because of the impossibility of controlling the action of the object on the measuring instruments if these are to serve their purpose—the necessity of a final renunciation of the classical ideal of causality and a radical revision of our attitude towards the problem of physical reality.[16]

This "radical revision of our attitude towards the problem of physical reality" does not consist of denying altogether any physical reality, such as Planck's quantum of action or the interaction between the quantum phenomena and the apparatus, but rather of how this is to be interpreted.

In particular, Bohr denied Einstein's definition of physical reality as consisting of independent physical properties inherent in the quantal object *prior* to its interacting with the apparatus or measuring device. In fact, he chided Einstein for maintaining this position as regards quantum mechanics when his own relativity theory had disconfirmed it for measurements of length, time, and mass. Moreover, Bohr's "principle of complementarity" was not a *denial* of physical reality, as Fine claims, but an attempt to conceptualize the fact that a complete quantum mechanical description of the experimental situation must include such seemingly contradictory properties as those of a particle and a wave, as these are disclosed under different experimental conditions. By taking into account the different experimental arrangements and the indeterminacy principle of Heisenberg, Bohr tried to show that the contradiction could be eliminated:

> ... there is essentially the question of *an influence on the very conditions which define the possible types of predictions regarding the future behavior of the system.* Since these conditions constitute an inherent element of the description of any phenomenon to which the term "physical reality" can be properly attached, we see that the argumentation of the mentioned authors does not justify their conclusion....In fact, it is only the mutual exclusion of any two experimental procedures, permitting the unambiguous definition of

complementary physical quantities, which provides room for new physical laws, the coexistence of which might at first sight appear irreconcilable with the basic principles of science. It is just this entirely new situation as regards the description of physical phenomena, that the notion of *complementarity* aims at characterizing (*ibid.,* pp. 138–139).

Indeed, it would be unlikely that the scientist who was driven to explain the stability of the Rutherford atom, and the emission spectra of hydrogen in terms of the transition of electrons from different energy levels, would be "an archenemy of realism."

One of the clearest statements of Bohr's commitment to realism occurs in the article he wrote in the commemorative volume to Einstein edited by Paul A. Schilpp.

The preceding years had seen great progress in quantum physics due to a number of fundamental discoveries regarding the constitution and properties of atomic nuclei as well as due to important developments of the mathematical formalism taking the requirements of relativity theory into account. In the last respect, Dirac's ingenious quantum theory of the electron offered a most striking illustration of the power and fertility of the general quantum-mechanical way of description. In the phenomena of creation and annihilation of electron pairs we have in fact to do with new fundamental features of atomicity, which are intimately connected with the non-classical aspects of quantum statistics expressed in the exclusion principle, and which have demanded a still more far–reaching renunciation of explanation in terms of pictorial representation.[17]

If we are to take this statement of Bohr seriously, it does not imply non-realism, but just the opposite. Furthermore, Bohr does not renounce explanations as such, but only those in terms of "a pictorial representation"—the crux of his controversy with Einstein that Fine misrepresents.

Even Fine's interpretation of Heisenberg is misleading (cf. p. 92). While in a particular article one may adopt stringent methodological procedures, such as restricting oneself to observable measurable quantities, as when Heisenberg developed matrix mechanics, this may not indicate one's general position. For example, when Heisenberg explicitly describes his philosophy of physics, he says:

It should be noticed at this point that the Copenhagen interpretation of quantum theory is in no way positivistic. For, whereas positivism is based on the sensual perceptions of the observer as the elements of

reality [as in Mach's positivism], the Copenhagen interpretation regards things and processes which are describable in terms of classical
concepts, i.e., the actual, as the foundation of any physical interpretation.[18] (Brackets added.)

In fairness, it must be pointed out that this makes no claim about
the reality of the quantum domain as such. However, Heisenberg
also says of Bohr's principle of complementarity which incorporates
"mutually exclusive" pictures, that by "playing with both pictures. . .we finally get the right impression of the strange kind of
reality behind our atomic experiments" (*ibid.*, p. 49). Thus he does
not deny a reality underlying the quantum mechanical experiments,
but does emphasize its strangeness.

What I find most incongruous about Fine's article is that after all
his fulminations and remonstrations against realism ("[r]ealism is
dead" despite attempts "to pump up the ghostly shell and to give it
new life," for it "is well and truly dead, an we have work to get on
with, in identifying a suitable successor" [pp. 83–84], he ends up
endorsing a somewhat realistic position, albeit a minimalist one.
First he says, "I want to sketch out what seems to me a viable
nonrealist position, one that is slowly gathering support and that
seems a decent philosophy for postrealist times" (p. 84). However,
when describing his own position, the "natural ontological attitude"
(NOA), he asserts that "the core position is neither realist nor
antirealist; it mediates between the two" (p. 98). Yet previously he
had made the following statement:

> . . . if the scientists tell me that there really are molecules, and atoms,
> and ψ/J particles and, who knows, maybe even quarks, then so be it. I
> trust them and, thus, must accept that there really are such things,
> with their attendant properties and relations. Moreover, if the instru-
> mentalist. . .comes along to say that these entities, and their atten-
> dants, are just fictions (or the like), then I see no more reason to
> believe him than to believe that *he* is a fiction, made up (somehow)
> to do a job on me; which I do not believe. It seems, then, that I had
> better be a realist. One can summarize this homely and compelling
> line as follows: it is possible to accept the evidence of one's senses,
> and to accept, *in the same way,* the confirmed results of science only
> for a realist; hence, I should be one (and so should you!) (p. 95).

I frankly do not know what to make of these seemingly contradictory statements: "realism is dead;" it is necessary to develop "a
reliable successor;" "I want to sketch. . .a viable nonrealist posi-

tion;" "I had better be a realist;" and that the "core position is neither realist nor antirealist." I have not deliberately quoted Fine out of context, but I do find these various assertions perplexing, to say the least! It is as if Fine is emotionally opposed to realism and yet finds it intellectually necessary. Though he denies the realist's thesis of scientific progression and accepts "the possibility that explanatory efficacy can be achieved without the explanatory hypothesis being true" (p. 100), he also says his "natural ontological attitude"

> sanctions ordinary referential semantics and commits us, via truth, to the existence of the individuals, properties, relations, processes, and so forth referred to by the scientific statements that we accept as true. Our belief in their existence will be just as strong (or weak), as our belief in the truth of the bit of science involved, and degrees of belief here, presumably, will be tutored by ordinary relations of confirmation and evidential support, subject to the usual scientific canons (p. 98).

This is an excellent statement of scientific realism, but one I am surprised to find Fine endorsing.

As Fine's statement indicates, realism is *implied* by the progressive achievements of science, not *an additional explanation* of these achievements. One cannot consistently accept the accomplishments of science on the one hand, and deny the truth of the theories and the actuality of the components by means of which these accomplishments are attained, on the other. This does not presuppose standing outside of the framework of science and matching it with the world, as Fine claims (cf. p. 9), but merely inferring what must be the case if one accepts the successes of science. Until the antirealist can show how the progressive achievements of science can occur *even though the theories are false,* and how the scientific discoveries underlying scientific explanations can be attained *even though what is discovered does not exist,* I must conclude that the antirealist believes that science has magical powers. It is revealing that at the end of his article Fine quotes with approval Mach, the physicist who did claim throughout his writings that atoms were "fictions," only reluctantly conceding the possibility of their existence as a result of Rutherford's experimental discoveries. Given the development of atomic fission and fusion, and the remarkable experimental discoveries in particle physics in the last several decades, I wonder what Mach would say were he alive today?

Larry Laudan's critique of realism in the same volume presents a more serious challenge because the arguments are more carefully marshalled and supported by numerous historical examples (although like my own references to recent scientific developments, his examples consist mainly of citations rather than analyzed elucidations). Thus an adequate reply would require an extensive analysis of the scientific theories he cites to determine what is implied by their having been successful at one time and then later disconfirmed or discarded. Does this mean, as he concludes, that such theories had no referential significance or truth value? Unfortunately, a full response is beyond the scope of this chapter, therefore I shall present only a general reply. Yet this may not be so regrettable since a fundamental disagreement would probably be the result in any case, due to our basically different points of view. Like the well-known ambiguous figures that can be seen either as a vase or as two facing silhouettes, or as a duck or a rabbit, our views of the history of science are two contrasting gestalts. Where Laudan sees (like Fine) mainly a series of false theories, I detect a significant progression of knowledge; where he observes disconfirmations and discontinuities, I see conceptual developments and refinements; whereas he looks in vain for any evidence of approximate truth in scientific theories, I find it incongruous that their success could occur without this; whereas he disavows any linkage between theoretical success and reference, I discern an obvious necessary connection, and so on.

From my perspective, much of Laudan's criticism of realism can be obviated (as was Fine's) if one does not attempt to *explain* the success of science in terms of scientific realism, but accepts realism as *implied by* the successes of science. It is not that reference and approximate truth explain the success of science, but rather, given the success of science, it is difficult (for me) to understand how this could occur if the theories of science were *entirely* false and *never* designated anything that actually exists in the world (other than observed phenomena). In my opinion, rather than denying reference and approximate truth, it is *necessary to explicate* what sense can be made of these concepts given that they are implied by the success of science. The main questions, then, are (1) whether in fact some scientific theories have been successful explanations of phenomena and whether there has been a progression in the development of these theories, (2) what constitutes successful and progressive explanations, and (3) how is this success and progression to be accounted for?

An affirmative answer to the first part of the question I take to be an historical fact, in that I do not see how it can be denied that our present knowledge of the structure of the universe is truer and more comprehensive than that of Anaximander, Ptolemy, or Newton; that our current conception of the atomic composition of matter is superior to that of Democritus, Newton, and Rutherford; that our explanation of the evolutionary process is more exact than that of Empedocles, Lamark, and Darwin; and that our understanding of the correlation of mental and brain processes is more accurate than that of Aristotle (who thought consciousness was located in the heart), Descartes (who thought the nervous system functioned as a hydraulic system), or Kant (who attributed all the integration of experience to the mind). If anyone denies this, then there is no point in further discussion.

Regarding the second and third parts of the question, as to what the success of science consists of and how this is to be accounted for, it seems to me that Laudan largely overlooks three crucial ingredients in the success of science. According to him,

> realists. . .are working with a largely *pragmatic* notion to be couched in terms of a theory's workability or applicability. On this account, we would say that a theory is successful if it makes substantially correct predictions, if it leads to efficacious interventions in the natural order, and if it passes a battery of standard tests. One would like to be able to be more specific about what success amounts to, but the lack of a coherent theory of confirmation makes further specificity very difficult.[19]

Yet we can be more specific!

The three consequences of scientific inquiry that Laudan barely mentions that others find particularly impressive are the following: (1) that scientific theories imply the existence of previously unknown entities (such as Neptune, neutrons, neutrinos, anti-matter, charmed quarks) that are later discovered; (2) the precision of the measurements of experimentally determined theoretical properties (such as mass, isotopic spin, charge, modes of decay); and (3) that powerful theories enlighten other areas of investigation leading to unexpected explanations, discoveries, and confirmations, as certainly has been true of the atomic-molecular theory in general, and of quantum mechanics in particular. And since these achievements pertain to what Laudan calls "the deep-structural dimensions of a theory" (p. 231), accepting the achievements while denying the dimensions seems to me to be incoherent—investing science with magical or "miraculous" (Putnam's term) powers.

Furthermore, I think that Laudan, as an instrumentalist, draws a sharper distinction between the directly testable or observable consequences of a theory, and its implications regarding deep structure, than is possible for contemporary theories. He says, for example,

> the realist needs a riposte to the prima facie plausible claim that there is no necessary connection between increasing the accuracy of our deep–structural characterizations of nature and improvements at the level of phenomenological explanations, predictions, and manipulations (p. 233).

But I do not find this claim to be "prima facie" at all! As I pointed out earlier, recent developments in the technology of photography and computer graphics have enabled scientists to 'see' molecular structures and atomic arrangements that previously were purely inferential. And the "reposte" of the realist is provided by the remarkable experimental advances in the twentieth century of such powerful theories as relativity and quantum mechanics. Moreover, these developments refute Fine's statement that "the last two generations of physical scientists turned their backs on realism and have managed, nevertheless, to do science successfully without it" (p. 83). I believe a careful review of the history even of quantum mechanics shows the contrary.

In Chapter IV I described how the early developments in atomic physics, the experiments by Rutherford disclosing the initial structure of the atomic nucleus and Bohr's explanation of the spectral emission of hydrogen in terms of fixed electron shells and quantum jumps, led to the Saturnian model of the atom. Thus began the discovery of the more basic subatomic particles and forces composing the atomic domain that has enlightened nearly every area of physical research. Impressed by the resemblance of Bohr's electron orbits and standing waves, the similarity of the equations of geometrical optics and classical mechanics, and guided by Einstein's association of wave frequency and photon energy, Louis de Broglie in 1924 proposed attributing a wave function to the energy of the electron, suggesting that the latter be thought of as an interior element of a guiding wave—a kind of "matter-wave." This bold conjecture of associating waves with material particles was confirmed in 1927 when Germer and Davisson succeeded in producing diffraction patterns of atoms. Building on the work of de Broglie, Schrödinger dropped the notion of the interior particle altogether, developing the famous equations of wave mechanics which interpreted Bohr's electronic orbits as multi-dimensional waves.

At about the same time, Heisenberg, restricting his theoretical considerations exclusively to the experimental data (namely, the possible energy states of Bohr's atom and the intensity and frequency of the spectral lines), eliminated entirely Bohr's model of electronic orbits, utilizing the relatively unknown mathematics of matrices to interpret and deduce the experimental phenomena. Surprisingly, despite the difference in mathematical treatment, matrix mechanics proved to be mathematically equivalent to Schrödinger's wave mechanics. This in turn was succeeded by Max Born's conception of "probability waves."

Following the development of non–relativistic quantum mechanics by Bohr, de Broglie, Heisenberg, Pauli, Schrödinger, and Born, quantum electrodynamics was largely the work of Paul Dirac, around 1930. A combination of classical electrodynamics, the field structure of relativity theory, and quantum mechanics, quantum electrodynamics is a limited field theory which interprets particles not as fundamental entities but as solutions of quantum field equations, and therefore as manifestations of interacting fields. The consequent relativistic quantum field theory was developed in its present form mainly by F. J. Dyson, Sin–Itro Tomonaga, Richard Feynman, and Julian Schwinger. The renormalized (where the infinite values of the equations of quantum field theory are reduced to finite magnitudes) quantum field theory of electrodynamics is now called "quantum electrodynamics," or QED.[20] According to Chris Quigg at the Fermi National Accelerator Laboratory (Fermilab), QED

is the most successful of physical theories. Using calculation methods developed in the 1940's by Richard P. Feynman and others, it has achieved predictions of enormous accuracy, such as the infinitesimal effect of the photons radiated and absorbed by the electron on the magnetic moment generated by the electron's innate spin. Moreover, QED's descriptions of the electromagnetic interaction have been verified over an extraordinary range of distances, varying from less than 10^{-18} meter to more than 10^8 meters.[21]

The success and precision of these predictions should have a sobering effect on those philosophers under the impression that scientific theories are no more than "likely stories," or even fairy tales.

Following the development of QED, there occurred remarkable discoveries of elementary particles by utilizing progressively more powerful particle accelerators. Particle or high energy research consists of the transmutation of subatomic elements producing new particles with unforeseen properties. These particles, successively predicted by theorists such as Pauli, Dirac, Wigner, Yukawa, Gell-Mann, Schwinger Weinberg, and Glashow, are created out of mass-energy transformations according to Einstein's formula $E = MC^2$, and were later experimentally confirmed. Such a plethora of particles composing the nucleons and complementing the electron has been discovered, that a new kind of periodic table of elements has been formed composed of hadrons and leptons, with the former further divided into baryons and mesons. The leptons are irreducible, point-like particles consisting of the electron, muon, and tau, along with their three neutrino adjuncts. The hadrons are composite, consisting not only of the more familiar proton and neutron but of such recently discovered particles as the pion, kaon, and eta mesons, and the lambda, sigma, and other baryons. This classification of particles is based on the traditional properties of mass and spin, along with such unusual new characteristics as isotopic spin, mean lifetime, and typical mode of decay.

Along with the discovery of this world of particles and their associated antiparticles with similar properties but opposite charge, two new forces in addition to those of electromagnetism and gravity were posited: a strong force binding the nucleons (the protons and neutrons), and a weak force controlling radioactivity within the nucleus. Unlike gravity and electromagnetism, the two new forces are effective only within extremely short distances. Moreover, the traditional notion of forces exerting an influence at a distance was superseded by the conception of "interaction" among the classes of particles due to an exchange of 'virtual' particles, electromagnetism consisting of an exchange of photons, strong interactions an exchange of gluons, and weak interactions an exchange of vector bosons composed of W^-, W^+, and Z° particles. While leptons respond only to the weak force consisting of the exchange of photons and the three bosons, the hadrons react to both strong and weak forces.

For a while it was thought that the leptons and hadrons (in addition to photons and postulated gravitons) exhausted the basic classification of particles, but with the proliferation of the hadrons (the baryons and mesons)—approximately 200 particles were listed in 1984—it seemed natural to search for a still simpler substructure of these particles. So in 1964 Murray Gell-Mann and George Zweig independently proposed theories accounting for the composition of the hadrons. As is well known, Gell-Mann whimsically named these particles "quarks" from a passage in Joyce's *Finnegan's Wake* ("Three quarks for Muster Mark!"), and the term caught on.

In experiments analogous in design and consequence to the initial discoveries of Rutherford, high-energy electrons were directed at protons and neutrons with the different angles and energies of their scattering indicating that "some were colliding with pointlike, electrically charged objects within the protons and neutrons."[22] From these experimental results and interpretations, in conjunction with theoretical inferences to maintain the principles of symmetry applied to the hadronic interactions, such properties as spin, electric charge, color charge, and mass were assigned to the quarks. Unlike previous particles that always had integral charges, the quarks have the unusual feature of having fractional charges of plus two-thirds or minus one-third. Quarks never exist separately but are joined as pairs or triplets to form the hadrons, a quark and anti-quark pair composing a meson and three quarks making up a baryon. Although the theory of quarks was an entirely deductive or inferential model to account for the hadronic interactions, their possible combinations into hadrons has been experimentally verified.

Beginning with the property or charge of "strangeness" postulated by Gell-Mann, and later that of "charm" suggested by Glashow, physicists now account for all of the hadrons and their interactions in terms of combinations of quarks, classified under the following "flavors": up (u), down (d), charm (c), strange (s), bottom/beauty (b) and, to round out the classification, presumably top/truth (t). According to this fundamental scheme of classification, matter consists of the two groups of six leptons and six quarks, plus the elementary particle forces. Based on the Yang-Mills Gauge Theory (a mathematical theory that maintains symmetry under certain

variations of properties) and group theory developed by the mathe-
maticians Marius Sophus Lie and Elie-Joseph Cartan, a new quantum theory analogous to QED was constructed, baptized "quantum
chromodynamics," or QCD, by Gell-Mann. The name was appropriate because chromodynamics (unlike electrodynamics that was
based on electric charges), consists of a strong force or charge
called "color" that binds the quarks together within the hadrons.

However, not satisfied with two separate categories of basic
particles, the quarks and leptons, as well as three (ignoring the
graviton) independent exchange particles accounting for the electromagnetic, strong, and weak forces, Glashow, pursuing suggestions of his dissertation director, Julian Schwinger, decided to
search for a unified theory. The first stage in this unification was his
integration of weak and electromagnetic interactions into an "electroweak theory." Steven Weinberg, Glashow's classmate at Bronx
High School of Science, then at Cornell, added the important concept of symmetry breaking by means of the Higgs boson, thereby
explaining how Glashow's W and Z particles acquired mass, a crucial link in the search for unification. The following year Abdus
Salem contributed further to the explanation. Then in 1969 the
Dutch physicist 't Hooft showed in two brilliant papers how Cartan's group theory could be applied to gauge theories, enabling
Weinberg in 1973 to produce a successful gauge field theory of
strong forces involving the color charge. This was the basis of QCD.

But while the theory of quarks and QCD explained the strong
forces, they did not account for electroweak forces. So Glashow, in a
paper written with James Bjorken, proposed a new quark called
"charm" that provided a bridge between QCD and the electroweak
theory, adding theoretical support to both theories, and paving the
way for a grand unified theory. But the task of confirming the
existence of charm proved unusually difficult because quarks, as
indicated earlier, are never found in an independent state, existing
in self-enclosed couplets or triplets within the hadrons. Any attempt to isolate and confirm the existence of quarks results in a
new combination of quarks concealed in the hadrons. Finally, however, evidence of charmed particles was found by investigators in
Japan and at Brookhaven. Then Samuel Ting at Brookhaven and
Burton Richter at the Stanford Linear Accelerator (SLAC) confirmed
the existence of a new particle (called "J" by Ting and "psi" by
Richter) which is now known as the J/ψ particle. Considered to be a
new meson composed of a charmed quark and anti-quark, its dis-

covery confirmed the existence of charmed quarks. Ting and Richter received the Nobel Prize in 1976 for their discovery, and
Glashow, Weinberg, and Salem shared the Prize in 1979 "for their
contributions to the theory of the unified weak and electromagnetic interaction between elementary particles. . . .[22] To climax these
results, Carlo Rubbia and Simon van der Meer were awarded the
Nobel Prize in 1984 for their confirmation at CERN (the European
Nuclear Center near Geneva) of the existence of the W^-, W^+, and Z°
particles.

But the search for a grand unified theory combining both strong
and electroweak forces had barely begun.[23] A bold step towards
unification was made by Glashow and Georgi in a series of four
papers published in 1973 and 1974, one of which had the imposing
title, "Unity of All Elementary–Particle Forces." According to their
Grand Unified Theory (GUT), quarks and leptons would be combined into one family on the grounds that they decay into one
another. In addition, a "superweak force" was proposed unifying
strong and electroweak interactions. Later in 1974 Georgi wrote a
paper with Weinberg and Helen Quinn claiming that while at low
energies the three forces remain distinct, at very high temperatures
or energies, such as existed at the birth of the universe with the Big
Bang, the three forces unite. Although not verified, at least the
framework of a Grand Unified Theory has been introduced. As
Crease and Mann summarize these developments:

> The result is a ladder of theories. Firmly on the bottom is SU(3) x
> SU(2) x U(1)[the SU stands for special unity group, based on Cartan's
> group theory], whose predictions have been confirmed ("to the point
> of boredom," Georgi says). . . .The W and Z particles were discovered
> at CERN. . .last year [1983], but the theory was so well established by
> then the event was—for theorists at least—an anticlimax. The GUTs
> proposed by Georgi and Glashaw and other physicists, which fully
> unite the strong, weak, and electromagnetic forces, are the next rung
> on the ladder. Although as yet unconfirmed, these theories are con-
> sidered compelling by most physicists. Finally, at the top of the
> ladder, in the theoretical stratosphere, are supersymmetry and its
> cousins, which are organized according to a principle somewhat
> different from SU(5), though, like that model, they put apparently
> different particles together in groups. Supersymmetry groups are
> large enough to include gravity, but are so speculative that many
> experimenters doubt that they can ever be tested.[24] (Brackets added.)

The striking experimental confirmations of the brilliant theorizing of the last several decades, which has led to a general theory

unifying what seems to some physicists to be the ultimate particles and forces of the universe, has left physicists in a highly excited and optimistic state. It appears as if the basic components of the universe have been discovered, so that a complete and final explanation of the cosmos is within reach. In words reminiscent of John Trowbridge, the head of the physics department at Harvard at the turn of the century, who discouraged students from pursuing graduate work in physics in the belief that everything of importance had been accomplished, Harold Fritzsch, a co–worker of Gell–Mann's who contributed to QCD, says:

> Today we can state unequivocally that the physics of the atom is understood. A few details need to be cleared up, but that is all. . . .Important and fascinating discoveries have been made in high–energy physics, especially since 1969, and today it appears that physicists are about to take the important step towards a complete understanding of matter.[25]

In addition, Crease and Mann report Glashow as saying about GUTs, that

> "I'm convinced the theory is sound. . . .the essential framework of the GUTs simply explains too many things for it to be fundamentally wrong[the basic assumption of realists]. We have the outlines of a full theory, and a few experimental results aren't going to change that [one cannot help recalling the "two dark clouds" on the horizon of physics at the end of the nineteenth century]. Besides, it's a mistake for theorists to take too seriously the first batch of experimental results."[26] (Brackets added.)

The last part of this statement was added presumably because there already are experimental phenomena reported by Rubbia that appear to be incompatible with the GUTs model. So Rubbia, along with Schwinger and Weinberg, are less optimistic that the search for the scientific grail is about to end. Nonetheless, whether ultimately true or not, these developments in particle physics have had a major impact on other areas of science and technology, such as astronomy and cosmology, genetics and molecular biology, chemistry and geology, and the technology of microchips and superconductors. These important consequences lend considerable additional weight to the belief that such theories must reflect *some* truth about the world, otherwise how can their effectiveness be explained? They also stand in marked contrast to Fine's claim quoted earlier, that "the last two generations of physical scientists turned their backs on realism."

I described these developments in particle physics in some detail since I think many philosophers hold the views they do about science because their familiarity with science is derived mainly through the writings of the British Empiricists, such as Hume. But the situation in physics today (as well as in chemistry, biology, and astronomy) is so unlike the state of physics in Hume's day that (as Galileo said of Aristotle) were he alive today, I believe even Hume would not subscribe to his original views of causality and induction. Furthermore, these changes in our capacity to experimentally disclose more of the microstructures of nature enable us to appreciate what lies behind the pragmatic criterion of truth. The fact that the states and occurrences of the world as directly encountered are increasingly explainable and controllable in reference to *experimentally discovered underlying structures and interactions* is what gave rise to and sanctions the pragmatic criterion. Not being able to test our hypotheses as to how or why things appear, occur, and react as they do by direct confrontation, we take an experimental or pragmatic approach, that of acting on one or more of the hypotheses so as to alter the conditions affecting the occurrences to determine whether the consequences of these experimental actions will confirm or disconfirm our original suppositions. Being pragmatic implies the willingness to act on various hypotheses to determine whether their implications regarding the resolution of a problem, when experimentally tested, prove to be the case or not. Thus when we cannot *ascertain in advance* of experimental testing whether our hypothetical solutions to a problem are true, we normally resort to the pragmatic criterion.

It is because the methodology of science has given man the capability of modifying and controlling, to some extent, the occurrences in the world around him that the pragmatists take as their paradigm of a thinker the scientific inquirer, the person confronted with some practical or theoretical problem whose resolution *has not yet* become a known fact or discovered or actualized state of affairs. Because the solution is not immediately determinable, one cannot ascertain *in advance* of any inquiry whether any suggested hypothesis, as expressed in a statement, corresponds to an existing or predicted state of affairs. In fact, in cases where we normally apply the pragmatic criterion of truth, the confirming or disconfirming state of affairs *does not even exist* until someone implements one of the possible hypotheses to determine or actualize its consequences. Thus if we are not sure which octane gasoline will perform best in our car, which flies will appeal to the trout that are biting, whether certain spices will enhance the flavor of a dish, how long something should be cooked to maximize its flavor and tenderness, if a particular selection of readings will prove the most effective in a course, or whether various curricular changes will bring about certain consequences, we utilize the pragmatic criterion to test which choices will have the desired effects. Precisely because the eventualities that will confirm our choices *depend upon conditions that do not exist* when the original questions are posed, we apply the pragmatic criterion so often in life.

This is especially evident as regards complex political, economic, and social problems. Consider, for example, the momentous problematic situations previously confronted: How can the United States win (or "end with honor") the Vietnam War?; What are the correct economic steps to take to alleviate a recession, reduce unemployment, and diminish inflation?; What are the best procedures for reducing our country's dependence upon foreign oil without, at the same time, disrupting the economy, causing undue hardship to specific segments of the society, or further desecrating the environment? In each of these cases, one cannot know *prior* to taking certain actions whether they will eventuate in the desired consequences because the confirming state of affairs not only does not exist when the question of the truth of a particular solution is raised, but *its very existence* is contingent upon many interrelated factors, one of which is the particular procedure or procedures one takes (the hypotheses one decides to act on) in dealing with the problem. As regards economic problems, for example, the mere announcement of a particular policy will itself affect the economic situation. *It is not until after one has acted on certain hypotheses that the situation will eventuate which will either confirm or disconfirm the proposed solution.* Obvious as this is, it has not been sufficiently emphasized even by the pragmatists themselves!

A clear illustration is the tragic history of the Vietnam War. I believe few people in the United States today would say, given the now evident consequences, that the original decision to enter the war (for whatever reasons) was correct. Thus, though one might have believed initially that the intervention on behalf of South Vietnam was justified (considering the plights of Poland, Hungary, and Czechoslovakia where we did not intervene), the subsequent consequences of intervention proved so disastrous, not only to the harmony and self-esteem of the people of the United States, but especially to the people of South Vietnam and Cambodia that, in terms of *these* consequences, hardly anyone would now claim it was a correct decision to have made.

In such complex examples, one can argue, as a number of people have, that it was not the *original decision* that was proven false, but the (limited) military operation taken to implement it ("we could have blown North Vietnam off the map"). Yet no one can be sure of the consequences of such an alternative, drastic action: for example, the lasting effects on world opinion toward the United States which cherishes the conviction that it tries to act morally; whether such an action would have led to the intervention of China and/or Russia; whether this in turn would have led to a nuclear war, and so on. Moreover, those who were responsible for the original decision to intervene did not anticipate that such extreme military measures would be necessary: if they had, there would have been many more misgivings initially, and probably a different decision made. But prior to the Tonkin Bay Resolution which brought about an actual condition of war no one could be certain what the effects of our support of South Vietnam would prove to be.

Similarly, those economic advisers or forecasters who propose various ways of stimulating the economy, of reducing inflation, and of increasing employment cannot know *in advance* whether the consequences will prove them correct, since there are so many complex, interdependent factors involved. As in quantum mechanics, where the experimental conditions for determining the position of a particle with unlimited accuracy preclude determining the momentum with a similar precision (and vice versa), owing to the irreducible quantum of action (as well as the wave-particle duality), so it appears that such economic states as inflation, unemployment, interest rates, international trade balance, and value of the dollar are so interrelated that the conditions affecting and governing some of these factors preclude controlling the other to the desired degree. (This is also why, since it is an entirely new legislative act, we cannot know the ultimate effects of the Gramm-Rudman-Hollings budget-balancing law, and therefore can only judge its success pragmatically.)

But the pragmatic criterion of truth is not applicable only to social, economic, and political proposals where the conditions are obviously malleable and the outcome initially indeterminate. It applies also to scientific and/or technological hypotheses, such as the earlier decisions to develop the atomic bomb and to attempt a lunar landing, since one could not be certain at the outset if success could be attained. This is also true today regarding President Reagan's decision to go forward with the Stratetic Defense Initiative ("Star Wars"). And it was true of Watson and Crick's conjecture that inherited characteristics are carried and determined by the molecular structure of nucleic acids. Watson and Crick could not have known prior to their exciting discovery whether their initial conviction, and the procedures and conjectures subsequently undertaken, would lead to success. Unlike the examples of the atomic bomb, the Vietnam War, and the lunar landing, the confirming state of affairs (the actual DNA and RNA molecular structure) did *preexist* before the investigation of Watson and Crick; but since the test of their hypothesis did not depend upon an *initial conformity* of the hypothesis with an *already known* genetic–molecular structure, and since the final solution had to be gradually pieced together on the basis of such diverse activities as seeking out and collating the appropriate data, correctly interpreting X–ray diffraction patterns of relevant molecular structures, utilizing knowledge from chemistry and molecular biology, and constructing ingenious models, the initial *primary* criterion on which the truth of the hypothesis depended was the pragmatic criterion. This is true even though the *eventual* recognition of the correctness of their final model depended upon its conforming to objective facts and its being consistent with all the relevant data. As stated earlier, each criterion has its own *predominant* or *proper* application under certain conditions which does not preclude or exclude the fulfillment (to a lesser degree) of the other criteria. In fact, this is almost always the case.

For any hypothesis to be true, what it implies must conform to some objective fact or state of affairs (however conditional or even momentary these might be, such as the discovery of extremely short–lived decaying subatomic particles, or a crisis situation like a hijacking), and be coherent with other theoretical concepts and empirical data. *But if the determination of the truth of the hypothesis depends upon actions or experiments that bring about the eventual verifying conditions or states which, therefore, cannot be known in advance of such procedures, then one utilizes the pragmatic criterion.* This explains the significance of the frequently quoted definition of the pragmatic criterion that "an idea or belief is true if it works." This conception has often been criticized on the grounds that it makes truth "crass," "expedient," or "self-serving," but once we understand what is meant by "works," this criticism is largely prejudicial.

Since the pragmatist believes that curiosity and thought are usually aroused when we confront a problem whether practical, theoretical, or even imaginary, the test of the truth of the statement expressing the proposed solution is whether it *actually solves the problem*. Thus the statement, "an idea is true if it works," means that if, acting on the basis of the proposed idea, the resultant consequences resolve the intially troubling, uncertain, problematic situation, then the original assertion expressing the idea is true. But, as argued earlier, there is considerably more involved here than taken into account by the usual "instrumentalist" interpretation of most pragmatists. Emphasizing the tentative and transitory nature of so much of scientific knowledge and of truth claims, and wishing to avoid any ontological commitment regarding the status of the referents even of confirmed scientific theories, philosophers such as Dewey, Rorty, Laudan, and Fine have equated the significance of scientific knowledge entirely with its success in solving problems. They are willing to grant the predictive confirmations and practical consequences of science, but divest it of any ontological importance, since they deny reality to the kinds of structures and mechanisms utilized by scientists in their explanations. In my opinion, this thesis is completely incoherent, as I argued earlier.

Instrumentalism is no more adequate as an epistemological interpretation of scientific discoveries and explanations than the positivists' thesis of phenomenalism, or Bas C. van Fraassen's conception of "constructive empiricism."[27] While these views had more plausability in the early decades of the twentieth century when developments in physics, chemistry, and neurophysiology were still in an uncertain stage, this is no longer the case. As was true of phenomenalism, it is exceedingly anthropocentric to suppose that only those properties of the physical world that are the result of interactions with an organism like man, are real, denying the possibility of disclosing other properties either by experimentation or by highly advanced techniques of observation. It is ironic that while the emergence of pragmatism coincided with the changing methods of investigation and conception of the world owing to the progressive influence of the sciences, pragmatists generally have been unwilling to acknowledge the ultimate fruits of these developments, a disclosure of many more dimensions of physical reality than we had any inkling of previously. And though a certain skepticism regarding the claims of some scientists to be on the verge of unlocking nature's ultimate secrets is justified, one should not overlook the fact that many scientific theories or discoveries are unlikely ever to be disproven: Harvey's 'theory' of the circulation of the blood and conception of the heart as functioning like a pump; Lavoisier's 'theory' of combustion; the causes of malaria and polio; the atomic–molecular explanation of chemical reactions; Cajal's neuronal theory, and so forth. That we may never arrive at a complete understanding of the universe does not mean that some things cannot be explained now. It is a fortunate feature of the universe that we do not have to know everything to explain something, at least relative to certain conditions or contexts.

The test of truth as determined by the pragmatic criterion is more flexible than that of the correspondence or coherence criterion. And since the characteristics of what will resolve a problem vary with the nature of the problem, pragmatists have not attempted to provide general criteria as to how these are known, although in actual practice *the problem itself specifies or generates* (as inquiry proceeds) *the criteria of its resolution,* so that in the final stage (again, the Watson and Crick discovery serves as an excellent example), there is little difficulty in objectively recognizing whether the initial problem has been resolved or not. Moreover, this account of the pragmatic criterion also accommodates those cases wherein theories considered at one time to be correct solutions to problems, relative to the information available, are replaced by newer theories with greater explanatory power, as happens constantly in the sciences.

Although the use of this criterion implies that the test of truth occurs within certain contexts or under certain conditions, and thus is somewhat relative, it does not seem to me that this implication is either crass, expedient, or self–serving. Insofar as nearly all problems are intersubjectively identifiable, and the criteria of their solution (as generated by the problems themselves) subject to *intractable external constraints* with *publically accessible* evidence (as in the demands of scientists that experimental results be reproducible under similar conditions by other investigators), the application of this criterion is no more susceptible to subjective abuse—and is no less rigorous—than any other criterion. Its justification is simply inherent in the conditions of empirical inquiry.

This is particularly true of quantum mechanical investigations, where the experimental outcome seems to be intrinsically related to the way the experiment is conducted, so the observer, investigative procedure and apparatus, and results make up an inseparable context. Though not yet recognized, this is true not just of esoteric investigations in quantum mechanics, but also of the most ordinary experiences, such as visual perception. Thus contrary to appearances, *what features the world disclose seem to be an inseparable function of the interaction of the investigative apparatus* (be it a living organism or a constructed instrument) *with the world, so that any characterization of the world apart from the specification of these conditions is fundamentally artificial*—as natural and useful as it might be for most purposes. If, as I believe, this is true, then of all the modifications in our world view brought about by modern science, this is the most radical and basic by far, an alteration even more unsettling in its implications than the Copernican Revolution. This revision in our world view will be described in the final chapter as "Contextual Realism."

Notes to Chapter VII

1. Cf. John Dewey, *The Quest for Certainty* (New York: G. P. Putnam's Sons, 1929), ch. II. The subsequent references to Dewey are to this work.
2. Cf. Richard H. Schlagel, *From Myth to the Modern Mind,* Vol. I, *Animism to Archimedes* (Bern: Peter Lang, 1985), ch. X.
3. Sir Isaac Newton, *Optics,* based on the 4th edition London, 1730 (New York: Dover Publications, 1952), p. 400.
4. Sir Isaac Newton, *Principia,* Vol. I, Motte's trans. revised by Cajori (Berkeley: University of California Press, 1962), p. 8.
5. Cf. A. Einstein, H. A. Lorentz, H. Weyl, H. Minkowski, *The Principle of Relativity,* trans. by W. Perrett and G. B. Jeffery (New York: Dover Publications, 1923), pp. 75-91; G. J. Whitrow, *The Structure and Evolution of the Universe* (New York: Harper & Brothers, 1959), ch. 4; and Milic Capek, *The Philosophical Foundations of Modern Science* (New York: D. Van Nostrand, 1961), Part II.
6. David Bohm, *Quantum Mechanics* (New York: Prentice-Hall, 1951), pp. iii–iv.
7. Cf. John von Neumann, *The Mathematical Foundations of Quantum Mechanics,* trans. by R. T. Beyer (Princeton: Princeton University Press, 1955), p. 253; Garrett Birkoff and John von Neumann, "The Logic of Quantum Mechanics," *Annals of Mechanics,* Vol. 37, 1936.
8. Cf. Henry Staff, "Are Superluminal Connections Necessary?", *Nuovo Cimento,* 1975, 29B, p. 271; Gary Zukav, *The Dancing Wu Li Masters* (New York: William Morrow, 1979), pp. 297-331.
9. Cf. David Bohm, *op. cit.,* pp. 612-622; Gary Zukav, *op. cit.,* pp. 304-331; and Richard Schlegel, *Superposition and Interaction* (Chicago: University of Chicago Press, 1980), ch. 6.
10. Cf. Bryce S. De Witt and Neill Graham (eds.), *The Many-Worlds Interpretation of Quantum Mechanics* (Princeton: Princeton University Press, 1973).
11. Cf. Fritjof Capra, *The Tao of Physics* (Boulder Co.: Shambhala Publications, 1975; David Bohm, *Wholeness And The Implicate Order* (London: Routledge & Kegan Paul, 1980), ch. 7.
12. Paul Churchland, *Scientific Realism and the Pasticity of Mind* (Cambridge: Cambridge University Press, 1979), p. 2.
13. Cf. C. I. Lewis, *Mind And The World Order* (New York: Dover Publications, 1929), pp. 266-273.

14. This summary is based on Charles Singer's, *A Short History of Scientific Ideas* (London: Oxford University Press, 1959), chs. XIII, IX.

15. Arthur Fine, "The Natural Ontological Attitude," *Scientific Realism,* ed. by Jarrett Leplin (Berkeley: University Press, 1984), pp. 83-107. All subsequent references to Fine are to this article.

16. Neils Bohr, "Can Quantum-Mechanical Description of Physical Reality be Considered Complete?", reprinted in *Physical Reality,* ed. by Stephen Toulmin (New York: Harper & Row, 1970), p. 132.

17. Niels Bohr, "Discussion with Einstein on Epistemological Problems in Atomic Physics," *Albert Einstein: Philosopher-Scientist,* ed. by Paul A. Schilpp (Evanston, IL.: Library of Living Philosophers, 1949), p. 237.

18. Werner Heisenberg, *Physics and Philosophy* (New York: Harper & Row, 1958), p. 145.

19. Larry Laudan, "A Confutation of Convergent Realism," *Scientific Realism,* ed. by Jarrett Leplin, *op. cit.,* p. 222.

20. I am indebted in much of the following discussion to the excellent article by Robert P. Crease and Charles C. Mann, "How The Universe Works," *The Atlantic Monthly,* August, 1984, pp. 66-93.

21. Chris Quigg, "Elementary Particles and Forces," *Scientific American,* April, 1985, p. 89.

22. *Ibid.,* p. 86.

23. Cf. Robert P. Crease and Charles C. Mann, *op. cit.,* p. 88.

24. *Ibid.,* p. 91.

25. Harold Fritzsch, *Quarks,* trans. by M. Roloff and the author (New York: Basic Books, 1983), p. 10.

26. Robert P. Crease and Charles C. Mann, *op. cit.,* p. 92.

27. Cf. Bas C. van Fraassen, *The Scientific Image* (Oxford: Clarendon Press, 1980). A separate article is being written on van Fraassen's book, so I have not discussed it here.

CHAPTER VIII

THE COHERENCE CRITERION OF TRUTH

THROUGHOUT the previous discussion the indispensability of concepts and language for the acquisition of knowledge—that all knowledge is framework dependent—has constantly been emphasized, yet in our examination of the correspondence and the pragmatic criteria of truth, the crucial function of a framework of interpretation was not fully brought out and discussed. What the application of both these criteria presupposes, and what the coherence criterion takes into account, is that the "correspondence" of a statement with what it designates or the "pragmatic" resolution of a problematic situation presupposes a theoretical framework in terms of which the confirmation or resolution takes place. As the identification of a state of affairs and the recognition of a resolved problem depend upon a background conceptual–linguistic system, any test of truth is not only a test of a particular statement, but is also an indirect affirmation of the framework within which the particular statement is confirmed. As C. I. Lewis maintained:

> We cannot even interrogate experience without a network of categories and definitive concepts. Until our meanings are definite and our classifications are fixed, experience cannot conceivably determine

anything. We must first be in possession of criteria which tell us what experience would answer what questions, and how, before observation or experiment could tell us anything.[1]

Moreover, if we reject the positivists' distinction between theoretical and observation statements, along with their belief that the latter are epistemically privileged (in the sense that they are not liable to falsification because they are matched to direct observations), then coherence has to be our ultimate criterion of truth. As I argued in Chapter V in opposition to Quine, if even the most direct reports of our peripheral sensory stimulations can be articulated in verbally different ways, and if even these basic statements can be falsified as a result of scientific developments (for example, "the sun rises and sets," "the earth is stationary," and "I am seeing the stars as they exist now"), then observation statements lose their foundational status in knowledge. If all statements are theoretical and susceptible to falsification, none can be selected as unconditionally true because of its transparent agreement with reality. As I argued in Chapter VI, for everyday occurrences we act as if our use of ordinary language mirrors reality, but this is because natural languages and ordinary experience developed together. With the growth of science and its increasing effects on our beliefs, we now realize that the world depicted in our ordinary languages is only a superficial representation of reality. As Berkeley asserted, we "speak with the vulgar but think with the learned." And if no statement by itself "wears its evidence on its sleeve," then the final justification of truth must depend upon *the most coherent integration of knowledge.* The contention that because all knowledge is theoretical we should assign greater significance to our scientific theories than to our commonplace observations or judgments, has been forcefully advanced by Churchland:

> As with singular theoretical judgements generally then, the adequacy of our perceptual judgements is in part a matter of the adequacy of the background theory (conceptual framework) in whose terms they happen to be framed. Our perceptual judgements can no longer be assigned any privileged status as independent and theory–neutral arbiters of what there is in the world. Excellence of theory emerges as the fundamental measure of all ontology. The function of science, therefore, is to provide us with a superior and (in the long run) perhaps profoundly different conception of the world, *even at the perceptual level.*[2]

Accordingly, any assertion of truth or any proposed hypothesis guiding inquiry which is subsequently confirmed or disconfirmed contains certain theoretical assumptions, conceptual interpretations, and linguistic or symbolic conventions. If one encounters phenomena which do not conform to the expectations predicted on the basis of the framework, there is the difficult problem of deciding whether the data (often initially amibiguous in any case) should be accepted and the framework revised accordingly, or whether the data themselves should be rejected as "unreal," "illusions of the senses," "artificially produced by the apparatus," or "otherwise explainable." Historical examples are the best illustration of this complex interdependence of theoretical presuppositions and 'observable facts.'

In April, 1610, armed with his perfected telescope and supreme confidence, Galileo visited the University of Bologna to convince his major scientific opponents, led by Giovanni Antonio Magini, of his marvelous telescopic discoveries. However, when his critics attempted to duplicate and verify his observations, the outcome was anything but satisfactory, as a witness describes in a letter to Kepler.

"Galileo Galilei, Paduan mathematician, came to us at Bologna, bringing his telescope with which he saw four *feigned* planets. I never slept on the twenty-fourth or twenty-fifth of April, day or night, but I tested this instrument of Galileo's in a thousand ways, both on things here below and on those above. Below, it works wonderfully; in the sky it *deceives* one, as some fixed stars are seen double. I have as witnesses most excellent men and noble doctors...and all have admitted the instrument to *deceive*. Galileo fell speechless, and on the twenty-sixth...departed sadly from the distinguished Doctor Magini."[3] (Italics added.)

Thus while Galileo was convinced of the validity of his observations, at first they were rejected by his critics as "illusions of the lenses," a dramatic example of the difficulty one can encounter in evaluating the evidence supporting a new theory. For as numerous scholars have pointed out, without a valid optical explanation as to how the lenses in a telescope functioned (an explanation that Galileo claimed to possess but which actually was provided later by Kepler), it was not initially evident whether Galileo's telescopic observations were in fact evidence of new astronomical discoveries, or whether they were due to optical distortions in the lenses. Although it could be shown that the telescope functioned reliably as far as terrestrial observations were concerned, it was not certain whether this was true of astronomical observations as well. (Because of such ambiguous interpretations and the contentious nature of the supporting arguments, Feyerabend uses this and similar examples to justify his provocative view that "[s]cience is an essentially anarchistic enterprise...."[4]).

By the following December, however, Galileo succeeded in convincing his scientific opponents, including Magini, of the genuineness of his observations (indicating that the confirmation situation is not as capricious as Feyerabend believes), but there was still the problem of the Church's resistance. Having persuaded the major scientific skeptics of the authenticity of his telescopic discoveries, Galileo offered them as evidence for the Copernican theory. However, his adversaries within the Church, among them Cardinal Bellarmine, in order to forestall too radical an assault on the traditional cosmological–theological framework, and because the evidence was by no means conclusive, tried to persuade him to admit that while the Copernican view might be a more *convenient astronomical hypothesis,* it was not a *true description* of the universe.[5] But this suggestion ran counter to Galileo's uncompromising conviction that the heliocentric theory represented a *true description* of the solar system (a fact that Fine and Laudan should note). Eventually, as we know, Galileo was vindicated, his astronomical evidence for and theoretical arguments in defense of the Copernican theory finally accepted—although, ironically, the argument that he was convinced was *the proof* of the heliocentric view, that the tides were caused by the axial rotation of the earth, was false. Yet the tragic affair illustrates the polemics that can occur at any time regarding how much weight to attach to the empirical data (or even as to what the empirical data mean), in contrast to upsetting the traditional theoretical framework, and the kinds of compromises that can be advanced.

The initial reaction to the 'null effects' of the Michelson–Morley (hereafter referred to as M-M) experiments, referred to previously, is another excellent example. On the basis of the Galilean principle of the addition of velocities, it was predicted that the velocity of light would be affected by the motion of the earth through the ether, and thus would have a velocity dependent upon how it was measured in relation to the direction of the earth's motion. But regardless of when or where in the earth's orbital trajectory the experiment was conducted, the velocity of light was found to be constant. To appreciate the unexpected import of this, imagine measuring the speed of light from a position of relative rest and then accelerating to nearly the same velocity as light and finding that its velocity (relative to one's own) was *still the same*! Given this astounding experimental consequence, some kind of accommodation had to be made.

The simplest reaction would have been to find some fault with the experimental procedure itself, thus discrediting the experimental results. However, as the experiments were conducted independently by both Michelson and Morley under the most exacting conditions, with the same consequences, this approach proved futile.[6] So it became a question of revising one or another of the fundamental presuppositions or concepts on the basis of which the experiment was conducted. It has been pointed out, somewhat facetiously, that if the experiment could have been performed during the height of the Copernican controversy it would have been interpreted as a "crucial experiment" proving the truth of the Ptolemaic view that the earth was stationary. For one of the tenets underlying the M–M experiments was the 'fact' that the earth revolves in its orbit at a rate of twenty miles per second, so that the measured velocity of light would be increased or decreased depending upon whether the light beam was projected in the same direction as the earth's motion or perpendicular to it. But if the earth were *stationary* it would not affect the velocity of light, thus explaining the null effects of the M–M experiments.

Another crucial assumption was the existence of the ether, held by almost all scientists at the time to be essential for the propagation of electromagnetic waves, and thus considered one of the fundamental components of reality. As Oliver Heaviside declared, ether and energy were the only true realities, everything else was "moonshine." Nevertheless, since the changes in the velocity of light were believed to be due to an "ether–wind" effect caused by the motion of the earth through the ether, the unexpected results of the M–M experiments also could be accounted for by abandoning the concept of the ether, as incredible as this appeared at the time. But there were other interpretations even more startling.

In the actual interferometer experiment, a beam of light was split and sent in perpendicular directions along arms of equal length of the interferometer, at the end of which they were reflected by mirrors back to their origin. Because of their perpendicular trajectories it was assumed that if one beam *traversed* the path of the ether on its round–trip trajectory, the other would necessarily move *with* the ether in half its course and *against* it in the other half (imagine a river with a current on which two boats are propelled at the same rate, beginning at the same point and moving the same distance, but one across the river and back, the other along the river and back). Since simple mechanical principles predicted that the latter round trip would take slightly longer than the former, the synchronous return of the light beams could be accounted for by assuming that in measuring the length of the arm of the interferometer facing the direction of the ether–wind, the measuring rod *contracted* just the amount necessary for the simultaneous return of both beams, an explanation proposed by G. F FitzGerald. Thus the null effects of the M–M experiments could be explained by assuming that movement in the direction of the ether would result in a contraction of measuring rods (Lorentz later proposed an analogous retardation of clocks)—an assumption that, however, opposed the age–old conviction that spatial extension and temporal durations were *intrinsic metrics* impervious to alteration due to velocity (or the strength of a gravitational field).

But while FitzGerald and Lorentz introduced these hypotheses in an *ad hoc* manner to account for the *apparent* constancy of the velocity of light indicated by the M–M experiments, Einstein's investigaton of certain anomalies in electrodynamics had already led him to conclude that the velocity of light was constant, and therefore that measurements of *the same event* by observers in relative motion must result in their attributing *different distances and times* to the event, relative to their own motions.[7] Thus the classic assumption that space and time were absolute, independent metrics underlying all measurements, so that all observers could arrive at an identical (instantaneous) value for an event, was operationally disconfirmed and invalidated (this consequence, incidentally, was one of the main reasons the positivists adopted the "operational" and "verifiability" theories of meaning).

As this example illustrates, profound theoretical revisions can occur as a result of encountering experimental data that are not in accord with the traditional framework, even though their discovery was made on the basis of traditional assumptions (Planck's investigation of blackbody radiation leading him to introduce a discontinuous physical magnitude, the quantum of energy, would be another example). As the histories of Galileo and the Michelson–Morley experiments illustrate, however firmly entrenched certain traditional frameworks might be, nature can always impose new experimental results that cannot be assimilated into them. As argued previously, this ensures that *while all knowledge is framework dependent, it also represents some accommodation to an objective reality.* Insofar as all empirical *observations* (especially when obtained by the use of highly complex and technical instruments) depend largely upon the presuppositions and expectations of the observer, and *any interpretation* of the experimental results depends upon many theoretical considerations, the *accommodation* of the framework to nature is balanced by the necessary *assimilation* of nature within some conceptual–linguistic scheme. As Quine states:

> Statements of the existence of various sorts of subvisible particles tend to be theoretical...and...are scarcely to be judged otherwise than by *coherence,* or by considerations of overall simplicity of a theory whose ultimate contacts with experience are remote as can be from the statements in question.[8] (Italics added.)

Thus not merely conformity to fact enters into the evaluation of observations and conceptual interpretations, but also such theoretical considerations as (1) the facility of the intrepretation (does it require the least conceptual readjustment?), (2) the simplicity of the interpretation (does it make the fewest number of independent assumptions?), (3) the resultant internal consistency of the framework, (4) the comprehensiveness of the interpretation (in reducing previously independent concepts, laws, or theories to a more unified system), (5) its testability (the feasibility of deriving decisive or unambiguous tests of the truth or falsity of the theory), and (6) its fruitfulness (the extent to which it leads to additional predictions and discoveries). How these different considerations are weighed depends upon the circumstances and the preferences and ingenuity of the scientist.

As Quine indicated, insofar as the truth of a theory depends upon the interrelation of these diverse theoretical factors, one is justified in claiming that the criterion used is that of coherence. Specifically, when the test of truth is not simply the direct recognition of states of affairs (correspondence) or of predicted consequences (pragmatism) *in terms of an established framework,* but depends upon the assimilation of new evidence, discoveries, or experimental results within an older framework that may have to be either revised or rejected to accommodate the newly acquired or reinterpreted data, then the total *coherence* of the interpretation becomes the primary factor in assessing its truth. Observable data, predictable consequences, and experimental results or discoveries are still essential, *but insofar as their meaning and significance depend upon how they are interpreted,* the coherence of the interpretation becomes the predominant factor in the assessment, even though the other criteria continue to play some role.

A clear example of the overriding significance of the coherence of the theoretical framework is illustrated in the recent defense of the identity or eliminative theory of the mind–body problem. Nothing would seem to be more transparently obvious or self-evident to the average person than the fact that he is aware of his own conscious states, whether mental musings, pains, or feelings of exhileration or depression, in a way that is inaccessible to anyone else (that our "priviledge access" to these experiences is known to us privately, in contrast to our shared knowledge of physical objects that can be publically observed). In addition, it is just as clearly evident that these conscious experiences are not neurological occurrences, especially as someone can be directly aware of the former and know nothing of the latter. This psycho-physical dualism is so enshrined in our common sense view of ourselves and embedded in our everyday speech that it seems absolutely incontrovertible.

Yet in order to promote a purely physicalistic or materialistic framework congruent with the physical sciences, identity or eliminative theorists maintain that this dualistic view of ourselves, dating back to earlier times, is merely a "folk psychology" and therefore can be reinterpreted if need be. Thus philosophers such as Place,[9] Smart,[10] Armstrong,[11] Feyerabend,[12] Churchland,[13] and Quine,[14] who advocate a unitary physicalistic or materialistic world view, and hence argue either for an elimination of the mental or for the reduction of consciousness to neurological processes, do not acknowledge any unanalyzable, irreducible conscious experiences that stand in the way of such a drastic reinterpretation. Whatever the apparent phenomenological 'facts,' whether of immediacy, privacy, or indubitability, these philosophers maintain they should be reanalyzed so as to be assimilated into a completely physicalistic world view. Armstrong, for example, claims that if we mean by 'mental,' 'what causes certain behavior,' then the mental can be equated with our brain, *tout court,*[15] while Feyerabend argues that the fact that the traditional mind-body distinction is enshrined in ordinary discourse merely points up the need for changing our linguistic conventions: rather than sanctifying indubitable facts, ordinary idioms may reflect a cultural or intellectual lag. As he says,

> there is the *fact* of knowledge by acquaintance. This fact refutes materialism which would exclude such a fact. The attack [on the above position] consists in pointing out that although knowledge by acquaintance may be a fact...this fact is the result of certain peculiarities of the language spoken *and therefore alterable*. Materialism...recognizes the fact and suggests that it be altered. It therefore clearly cannot be refuted by a repetition of the fact.[16] (Brackets added.)

What this position amounts to is the denial that there are any privileged or irreducible facts at all! Because all so-called facts are only facts relative to an accepted world view and some idiom, they can always be challenged by modifying one's world view and changing the idiom. As Feyerabend's position is similar to the linguistic relativity hypothesis of Whorf and Quine, it too raises the question why "certain peculiarities of language" arose in the first place, and what criteria are to be used in recommending particular changes of idiom. However, these problems do not concern Feyerabend who advocates the proliferation of theories, and therefore modifying common linguistic uses that stand in the way. As he says:

> Many...facts are formulated in terms of the [common] idiom and therefore already prejudiced in its favour. Also there are many facts which are inaccessible, *for empirical reasons,* to a person speaking a certain idiom and which become accessible only if a different idiom is introduced. This being the case, the construction of alternative points of view and of alternative languages which radically differ from the established usage, far from precipitating confusion, *is a necessary part of the examination of this usuage* and must be carried out *before* a final judgment is made (*ibid.,* p. 145; brackets added).

That the physicalist position assumes the coherence criterion of truth is evident from the fact that the commitment to it is not based on any conclusive empirical discoveries (although the physicalists are influenced, obviously, by recent advances in neurophysiology), but on the belief in the ultimate soundness of a "unitary" physicalistic world view—in a theoretical framework which is congruent with that of the physical sciences so that any incompatible data must be eliminated. As Smart states, "sensations, states of consciousness, do seem to be the one sort of thing left outside the physicalist picture, and for various reasons [namely, simplicity and economy] I just cannot believe that this can be so."[17] This explains the curious fact that while the adherents of physicalism offer the 'theory' as an empirical hypothesis, the 'theory' strikes its opponents more as a traditional metaphysical view (the converse of Berkeley's), since the arguments of the physicalists are not based so much on any decisive falsifying evidence or predictions, as they are on eliminating conceptual and/or linguistic obstacles to the acceptance of their preferred framework.

The position of the physicalist is that our common sense conception of ourselves as conscious beings, in addition to neurological organisms, is a legacy of an outmoded *theory,* one that originated before we knew anything about the neurological processes underlying our conscious states. This older psychic dualism, embedded in our conventional ways of talking about ourselves, appears to describe what is manifestly and incontrovertibly true of ourselves, while in fact it is only one possible conceptual–linguistic system that should be replaced by a better one in accordance with our current scientific discoveries in neurophysiology. As Churchland states:

> The eliminative materialist holds that the P–theory [our ordinary conception of ourselves as persons], not to put too fine an edge on the matter, is a *false* theory. Accordingly, when we finally manage to construct an adequate theory of our neurophysiological activity, that theory will simply displace its primitive precursor. The P–theory will be eliminated, as false theories are, and the familiar ontology of common–sense mental states will go the way of the Stoic pneumata, the alchemical essences, phlogiston, caloric, and the luminiferous aether.[18] (Brackets added.)

The underlying assumption of this position is *if* a completely physicalistic framework can be developed, it should be accepted, on the grounds that it is simpler and more coherent, given our current scientific theories. However, as was true of Berkeley's theory, when one urges the elimination of certain pervasive phenomena by extruding them from our theoretical framework because of certain *a priori* ontological commitments, there is always the danger of excluding an essential component of reality that may be indicative of the limitations of our current theories, rather than of the unreality of the phenomena. At one time one could have argued similarly for the elimination of magnetism, electricity, or radiation. Ernst Mach was a reductionist regarding physical theories such as atomism, whose view was completely refuted by later developments in atomic physics. Also, Einstein was criticized for pursuing a unified field theory that now may have been attained in the Grand Unified Theory—not by reduction!

The converse pertains to parapsychology. In contrast to accepting a theoretical framework and then attempting to reinterpret recalcitrant data to fit it, researchers in parapsychology have accumulated considerable empirical data which they accept as valid, but which cannot be explained scientifically—although interestingly enough, telepathy is consistent with the Einstein–Podolsky–Rosen (EPR) effect, which claims an "instantaneous" (or "telepathic") communication between separate quantum states, independent of any known physical causes.[19] Without begging the question, there seems to be some evidence for certain kinds of psychic experiences (telepathy, clairvoyance, psychokineses, and precognition) that cannot be explained by present scientific theories, but since there are no accepted alternative explanations of these experiences, they have had very little effect, so far, on our conception of man. Until we can provide some coherent explanation of the phenomena, either in a positive sense or by confirming that the data are spurious, they remain for most people curious but puzzling—just as, for over a century, the evidence for electricity and magnetism could not be incorporated into Newtonian science.

The influence of an accepted framework on the recognition and acceptance of certain phenomena is illustrated further in the curious problem of "ghosts." I have been told by a number of extremely objective, rational, "tough-minded" people (in England and Scotland), whose judgment I would unhesitatingly accept in other matters, of certain peculiar experiences (some of which they could report firsthand) which, *if authentic,* would necessitate a radical revision in our conception of man and "life after death." Nevertheless, I continue to think of ghosts as "unreal," and to question the "authenticity" of the evidence offered by these persons. Why is this so?

Surely it is not merely a question of my *not having seen* a ghost, since I believe in all sorts of things I have not seen, such as electric currents, light and sound waves, mesons, and the like. No, it is simply that the properties of ghosts (as characterized by people who have "encountered" them) do not conform to my categorical scheme. As Carnap stated:

> To recognize something as a real thing or event means to succeed in incorporating it into the framework of things. . .so that it fits together with the other things recognized as real, according to the rules of the framework.[20]

The reason that most of us today (in contrast to primitive peoples who readily accepted the existence of spirits, daemons, and the ghosts of dead ancestors) scoffingly dismiss ghosts as unreal is that they do not behave as do the physical objects around us (they cannot be publicly observed, they cannot be photographed, they have an anomalous space–time existence, and so forth), and their presence cannot be experimentally confirmed, unlike unobservable scientific entities.

But just as the "two small clouds" referred to by Lord Kelvin poised over the horizon of early twentieth century physics (the null results of the Michelson–Morley experiments and the failure of the Rayleigh–Jeans law to predict the radiant energy emitted by a black-body) burst into a torrent of revolutionary changes in physics in the early part of this century, so the anomalous phenomena of parapsychology, the recent investigations of auras with Kirlian photography, the current impasse in the mind–brain problem, and the paradoxical developments in quantum mechanics *could* lead to revolutionary conceptual changes as radical as any in the past. As many scholars have suggested, we may be facing a revolutionary breakthrough that will not only result in a tremendous new resynthesis of knowledge (similar to the earlier syntheses of Aristotle and Newton), but that will eventuate in a conception of the universe much stranger than any that we have conceived in the past.

Even so, this development, however remarkable, will undoubtedly be just another stage in man's attempt to understand the seemingly inexhaustible dimensions and domains of the universe, not the attainment of a final truth. Although the assumption that man can eventually acquire absolute knowledge has been one of the most persistent "idols" of mankind, history has not confirmed this expectation. Not only should we anticipate—indeed, welcome— equally extraordinary discoveries in the future, it might well be the case, as Kant and some leading quantum physicists have maintained, that the answers to man's fundamental questions lie beyond any knowledge attainable within the limits of the human condition, or even lie beyond his capacity to understand. Although Einstein asserted that "the most incomprehensible thing about the world is that it is comprehensible,"[21] this does not mean necessarily that the universe is completely or finally intelligible. In contrast to Einstein, Heisenberg posed the question to himself after a late night discussion with Bohr: "Can nature possibly be as absurd as it seemed to us in these atomic experiments?"[22]

To illustrate further the difficulty, at times, of isolating and evaluating the empirical data in contrast to the conceptual framework or the mathematical formalism, let us consider the extremely complex but fascinating case of quantum mechanics. The famous debate between Einstein (in collaboration with Podolsky and Rosen) and Niels Bohr, regarding the "incompleteness" or the "completeness" of quantum mechanics, indicates very clearly the reliance on different criteria of truth in the defense of their respective positions: Einstein evoking the *correspondence* criterion as the test of truth of any physical theory, with Bohr appealing to *coherence* as the criterion of truth of quantum mechanics. To avoid excessive complexity, I shall try to make the underlying principles, assumptions, and conceptual interpretations of the controversy clear to those unfamiliar with the more technical aspects.

The essence of the controversy pertains to the fact that while quantum mechanics contains a consistent formalism, within which the probable occurrence of "packets" or "ensembles" of subatomic events can be predicted with great accuracy, it does not predict individual events, and neither the formalism nor its conceptual interpretation provides an explanation as to the precise nature, development, and causal interaction of the quantum phenomena. For this reason, Einstein (along with de Broglie, Schrödinger, and Bohm) concluded that quantum mechanics represents an "incomplete" account of subatomic phenomena, while Bohr (along with Heisenberg, Born, and Dirac) maintained that the predictive success of quantum mechanics, albeit statistical, requires that we renounce the classical assumption of a *preexisting* infrastructure of determinant elements interacting according to exact causal laws, and accept a radically different conception of physical reality consisting of indeterminant, inscrutable, mutually exclusive but complementary processes — at least on the subatomic level which, however, is considered the basic level.

Thus the protagonists begin with their own definitions of the nature of physical theories and their respective criteria of truth (or "correctness"). Einstein–Podolsky–Rosen state their position as follows in their famous 1935 article:

> ANY SERIOUS consideration of a physical theory must take into account the distinction between the *objective reality,* which is *independent* of any theory, and the physical concepts with which the theory operates. These concepts are intended to *correspond* with the *objective reality,* and by means of these concepts we picture this reality to ourselves.[23] (Italics added.)

Having evoked the correspondence criterion of truth, the authors declare that the "success" of a theory will depend upon two factors: (1) whether the theory is "correct," and (2) whether the theory provides a "complete" description of the relevant phenomena, the latter depending upon whether *"every element of the physical reality"* has a *"counterpart in the physical theory"* (p. 124).

The answer to the first question depends upon the "correspondence" of the theory with physical reality, while the answer to the second rests upon an *exact determination* of the elements comprising the theory and the physical world. And since the relevant elements can be detected and defined only under experimental conditions, Einstein–Podolsky–Rosen implicitly appeal to the pragmatic criterion in specifying the nature of such elements.

> We shall be satisfied with the following criterion, which we regard as reasonable. *If, without in any way disturbing a system, we can predict with certainty. . .the value of a physical quantity, then there exists an element of physical reality corresponding to this physical quantity* (p. 124).

Finding certain quantifiable physical properties (such as mass, position, momentum, charge, wave frequency, etc.) indispensable for experimental inquiry, the authors conclude that such properties must have a counterpart in physical reality. But since in quantum mechanics the mathematical formalism only describes the possible or probable outcome of subatomic processes and leaves indeterminate the magnitudes of such properties as position, momentum, energy, and time, which are essential for the precise prediction of future states, quantum mechanics represents an "incomplete" description and explanation of microphysical processes.

In his reply in the following issue of the *Physical Review,* Bohr accused the authors of an "ambiguity" in the formulation of their criterion of physical reality in their expression "without in any way disturbing a system." This is the crux of the matter: whether the values of certain physical quantities in quantum mechanics can be ascertained without in any way disturbing the system. As Bohr states:

> Indeed the *finite interaction between object and measuring agencies* conditioned by the very existence of the quantum of action entails—because of the impossibility of controlling the action of the object on the measuring instruments if they are to serve their purpose—the necessity of a final renunciation of the classical ideal of causality and a radical revision of our attitude towards the problem of physical reality.[24]

According to Bohr, even if the account of subatomic phenomena given by quantum mechanics does not conform to the classical conception of physical reality, this "would hardly seem suited to affect the soundness of quantum-mechanical description, which is based on a coherent mathematical formalism covering automatically any procedure of measurement. . . ." (p. 131). Thus in contrast to Einstein-Podolsky-Rosen who hold to "correspondence" as the criterion of the truth of a physical system, Bohr claims that the "coherence" of the formalism with the experimental evidence is what constitutes its truth.

The fundamental assumption underlying the classical or Newtonian world picture, reasserted by Einstein-Podolsky-Rosen, was that whatever knowledge of quantifiable properties of one state of the universe was necessary to predict any subsequent state could, at least in principle, be experimentally obtained. If one knew the magnitude and direction of all the forces acting on a body and how they varied with time, along with the position and the velocity of the body at a specified moment, one could calculate precisely where the body would be found at any future moment. This was the basis of Laplace's famous assertion that "an intellect which at a given instant knew all the forces acting in nature, and the position of all things of which the world consists," could predict or retrodict all the motions of the universe from "the greatest bodies" to "the smallest atoms," so that "nothing would be *uncertain* for it. . . ."[25] But since such quantitative values as required by Laplace can, for human intellects, only be ascertained experimentally by using some detector instrument, the classical view also assumed (although the presupposition was so natural at the time that it was accepted as fact rather than assumption) that the instrument could be so used to render objective, absolute values of the necessary magnitudes, without in any way *disturbing* or *interfering* with the measured entities.

As natural as this assumption is, there are four interrelated developments in quantum mechanics which appear to make its fulfillment impossible: (1) Planck's discovery of a minimum quantum of action, (2) Heisenberg's uncertainty principle or principle of indeterminacy, (3) the wave–particle duality of electromagnetic and atomic phenomena, and (4) the fact that the basic equations or laws of quantum mechanics are statistical in form so that only probable occurrences can be predicted, implying an inherent element of indeterminacy and obscurity regarding the outcome of all subatomic processes.

Taking each of these developments in turn, in 1900 Planck published his momentous discovery that instead of the emission and absorption of energy occurring in continuous amounts, energy exchange occurs in discrete discontinuous quantities or quanta, the energy content of which is always an integral of $E = h\nu$, where E equals energy, h equals Planck's constant (calculated by Planck as 6.77×10^{-27} in c.g.s. units) and ν equals the frequency of the radiation. Instead of natural processes occurring in a continuous manner with differential coefficients that produce instantaneous or infinitesimal values necessary for exact calculations, occurrences at the subatomic level exhibit a granular structure with an irreducible quantum of action or interaction. This has both a physical and a mathematical significance: *physically,* it means that microphysical events consist of discontinuous exchanges of energy which prohibit the occurrences from having a vanishing magnitude; *mathematically,* it means that measurements on the quantum level cannot be reduced to infinitesimal magnitudes, introducing an unavoidable limit to the precision of quantum mechanical calculations analogous to the limit in precision that would be introduced into our measurements if we could only use integers and not fractional numbers. However, the latter limitation would be of only a notational nature, while the limitation introduced by Planck's constant is of a fundamental, physical nature manifested at a certain dimensional level of inquiry.

The second development pertains to Heisenberg's uncertainty principle originating with a *gedankenexperiment* concerning the problem of measuring the position and the momentum of sub-atomic particles such as electrons. The issue arises because the dimensions of the electron make it problematic as to whether its exact trajectory can be determined without interfering with it. While not a problem regarding macroscopic objects, measuring the properties of a subatomic particle is like trying to hold a fiery ember in one's finger to observe it more closely. Just as light waves are required to observe an ordinary object, to determine the position and the momentum of an electron (the quantities required to predict its future path) it is necessary to 'observe' the particle by interacting with it in some way. But this presents the following problem, as disclosed by Heisenberg.

In order to measure the *position* of the electron as precisely as possible, an extremely short wave length, such as a gamma ray, is required, otherwise the wave would bend around the electron. But as Einstein discovered, a very short wave length has a high frequency and therefore a large quantity of energy as indicated in Planck's formula $E = h\nu$ (where ν stands for the frequency of the radiation). This high frequency, large energy state required for the exact measurement of the *position* of the electron results in an inevitable and uncontrollable exchange of *momentum* between the gamma wave and the electron at the moment of interaction—this effect is called "uncontrollable" because it is uncertain whether the detection of the electron represents its position at the beginning, during, or immediately following the interaction. Thus the more precisely one determines the position the greater the exchange of momentum, introducing an inevitable "uncertainty" or "indeterminacy" in the measurement of the latter.

Conversely, to avoid affecting the *momentum* significantly would require the interacting wave to have a longer length with a lesser amount of energy, but this would result in an unavoidable imprecision in the measured *position*. So, while *either* the position *or* the momentum can be measured with an *unlimited accuracy,* the conjugate nature of the physical properties necessarily prevents an exact determination of both values. Letting delta p stand for the uncertainty in the measurement of the momentum and delta q the uncertainty in the measurement of the position, Heisenberg found that the product of the uncertainties must be equal to or greater than Planck's constant: $\Delta p \cdot \Delta q \cong h$. Thus the principle of indeterminacy states that the product of the uncertainty in the measurement of the position and the momentum, as well as the energy and the time, of subatomic particles cannot be reduced to a magnitude less than Planck's constant.

Expressed in the technical language of quantum mechanics, "canonically conjugate variables" such as those described above are not commutable: $p.q. \neq q.p.$ This means that in contrast to classical mechanics, where the "perturbation" due to the interaction of the phenomena with the detector instrument could be taken into account in the calculations, in quantum mechanics the manner and degree of the perturbations are, *in principle,* uncalculable. One has a "free choice" in arranging the experimental conditions to determine precisely either of the conjugate properties, but not both simultaneously, because the *experimental conditions* necessary to determine one of them *preclude* determination of the other. Thus the uncertainty or indeterminism is not due to a lack of knowledge on the experimenter's part that could possibly be removed, but is an inevitable result of investigations on the subatomic level. As Bohr states,

> we are not dealing with an incomplete description characterized by the arbitrary picking out of different elements of physical reality at the cost of sacrificing other elements, but....[with] the *impossibility,* in the field of quantum theory, of accurately controlling the reaction of the object on the measuring instruments....Indeed we have...not merely to do with an *ignorance* of the value of certain physical qualities, but with the *impossibility* of defining these quantities in an unambiguous way (pp. 136–137; brackets and italics added).

As interpreted by Bohr, this consequence means that in subatomic investigations one cannot make, as in classical physics, a definite distinction between the investigative conditions and the quantum phenomena. "Since these conditions constitute an inherent element of the description of any phenomenon to which the term 'physical reality' can be properly attached" (pp. 138–139), they form an inseparable unity. This explains why the experimental arrangements necessary for determining one of the pairs of conjugate properties exclude the other: *the properties are inseparable from the experimental conditions which, in turn, are mutually exclusive.* In answer to Einstein, Bohr argued that any attribution of "incompleteness" to quantum mechanics represents a misinterpretation of the theory because it presupposes the attainment of a kind of knowledge that, at the subatomic level, is not physically possible.

The third perplexing development involves the peculiar wave-particle duality of electromagnetic radiation and atomic particles. While certain kinds of optical processes (such as refraction and interference with diffraction patterns) are only interpretable in terms of wave configurations, other radiational phenomena (such as blackbody radiation and the photoelectric effect) are explainable only in terms of discrete particles or quanta. But since the properties of a particle and a wave are the obverse of each other—a particle having a precise localization in space and time, no two particles being able to occupy the same place at the same moment, and particles colliding and rebounding with a determinate energy exchange, while waves have a diffuse location tending to spread indefinitely in space, reinforcing or destructing when colliding at the same place, and preclude description in terms of energy exchanges resulting from collision and recoil[26]—it is difficult to understand why these mutually exclusive but complementary entities are both needed to explain electromagnetic phenomena.

Even though waves and particles have opposing properties, their manifestation under different experimental conditions prevents their being simply contradictory, as they would be if merely superimposed upon one another. Nevertheless, Planck's formula for determining the energy of a *particle,* $E = h\nu$, attributes the *wave property* of frequency to the *particle* while, by reflecting electrons off a crystal lattice, Davisson and Germer demonstrated that *electrons* manifest *diffraction patterns,* thereby confirming experimentally de Broglie's thesis that *all matter* has *wave properties.*

In addition, there are the baffling slit experiments in which successive single 'particles,' such as photons, are fired towards a diaphragm with double slits, one of which can be covered, and whose width can be regulated in proportion to the size (or wave length) of the photon. Behind the diaphragm a photosensitive screen registers where the individual photons strike. With *one* slit open sufficiently (diffraction patterns can also be produced with one opening smaller than the wave length), the photon passes through the slit and strikes the screen at a position reflecting its initial position and velocity. Theoretically, the photon could strike any area of the screen depending upon its trajectory. When the second slit is open, a diffraction pattern is produced on the screen, indicating that an interfering spherical wave emerged behind each openinig, the darkened lines of the diffraction pattern excluding the possibility of the photon striking the screen at that point. How can one explain these paradoxical results? With both slits open, one would expect the succession of individual photos to go through one slit or the other (depending upon their initial velocities), producing either the same pattern as with only one slit open, or two similar but smaller patterns. Why should the mere opening of a second slit change the emanation from the diaphragm from a particle pattern to two spherical waves? How does the particle 'know' the second slit has been opened and, if it does 'know,' how is the information conveyed? What is the nature of the interaction between the double slits and the photon that produces the wave effect? Why does the opening of the second slit preclude the photon from striking the interference areas? None of these questions can be answered in terms of current quantum theory.

Are photons, as de Broglie suggested regarding orbiting electrons, guided by "pilot waves" that normally are undetected but which under certain experimental conditions manifest themselves? Or, as Schrödinger conjectured, are all particles really "clouds of standing waves" which reduce to a kind of condensed singularity under suitable experimental conditions? Bohr attempted to mitigate the paradoxical implications of these experiments with the "principle of complementarity," maintaining that for a complete description of radiational phenomena both the particle and the wave aspects are necessary, though the experimental conditions under which either are manifested exclude each other.

> In quantum physics....evidence about atomic objects obtained by different experimental arrangements exhibits a novel kind of complementary relationship. Indeed, it must be recognized that such evidence which appears contradictory when combination into a single picture is attempted, exhausts all conceivable knowledge about the object. Far from restricting our efforts to put questions to nature in the form of experiments, the notion of *complementarity* simply characterizes the answers we can receive by such inquiry, whenever the interaction between the measuring instruments and the objects forms an integral part of the phenomena.[27]

The fourth development has to do with the expression of the basic laws of quantum mechanics in statistical form, which means that while the *probabilities* of all possible developments of the quantum phenomena can be precisely predicted, given the initial conditions, the actual outcome remains indeterminate until an interaction with some measuring apparatus occurs. The probabilities express deterministic laws as in classical physics, but unlike classical mechanics the actual occurrences are considered merely probable until the time of measurement. For example, Schrödinger's wave equation describes the causal or lawful development in isolation of the radiating phenomenon from its initial conditions to its interaction with a measuring system, but the wave function derived from the wave equation merely depicts the composite *possibilities* that *could* be actualized when the phenomenon interacts with some apparatus, not their actual existence prior to or at the time of measurement. The wave function describes the range of dynamic *possibilities* expressed in Schrödinger's wave equation, but does not describe *actual* developments of *existing* occurrences.

By squaring the amplitude of the wave function, $/\Psi/^2$, the probability density function is obtained, indicating the *probabilities* of each of the various *possibilities* represented by the wave function, but not which one of these will be actualized at the time of the measurement. As soon as the quantum phenomenon interacts with the detector instrument, all the possibilities depicted by the probability density function "collapse" into the one actuality that is registered by the apparatus. Which of the various probable possibilities is actualized is solely a matter of chance.[28]

Thus unlike classical mechanics where physical reality was assumed to have a precise nature and a continuous existence whether observed or not, and where interaction with the apparatus indicated a selection from previously existing properties with those non-selected still assumed to exist, in quantum mechanics the status of the phenomenon between measurements is merely probable, with all possibilities except one eliminated when the phenomenon interacts with the apparatus. It is as if the apparatus interferes with the possible developments in isolation of the quantum phenomena, precipitating an actual occurrence. According to Heisenberg, "generally there is no way of describing what happens between two consecutive observations."[29]

This concept of the probability wave was something entirely new in theoretical physics since Newton. . . . It introduced something standing in the middle between the idea of an event and the actual event, a strange kind of physical reality just in the middle between possibility and reality (*ibid.*, pp. 40–41).

While Einstein, Podolsky, and Rosen claimed that for any theory to be true it must be both "correct" and "complete," in the sense that *"every element of the physical reality must have a counterpart in the physical theory,"* in quantum mechanics the relevant elements have no determinate existence except at the moment of interaction with the measuring apparatus—prior to that they have only an indeterminate, probable status. Thus to insist on a "correspondence" between the elements of the theory and a prior physical reality, as if they were separable and independent, is to misinterpret the very nature of investigations on the subatomic level! Moreover, the previously cited statement presupposes that in some manner one can determine independently of the theory itself whether all the revelant elements of the physical reality are represented in the theory. But how can this be ascertained? Faced with this problem, Heisenberg reversed the question:

> Instead of asking: How can one in the known mathematical scheme express a given experimental situation? the other question was put: Is it true, perhaps, that only such experimental situations can arise in nature as can be expressed in the mathematical formalism (*ibid.*, p. 42)?

This means that rather than a theory being true because it correctly represents reality, *our conception of reality itself is dependent upon our theory.* Nothing could express more clearly the notion that all knowledge is framework dependent, with the *coherence* of the framework the ultimate test of truth. As Rorty states (in reference to the Quine–Sellars approach to epistemology), "there is no way to get outside our beliefs and our language so as to find some test [of truth] other than coherence."[30]

It was precisely this denial of independent, objective, determinant physical states and properties that Einstein, Podolsky, and Rosen hoped to disprove by their *gendankenexperiment.* While according to Heisenberg's principle of indeterminacy, the position and momentum of a particle could not be measured jointly with precision, the EPR thought experiment tried to show that one could *infer* from independent measurements of the initial positions and momenta of two interacting particles, the precise magnitude of these values at a later time, and thus prove that *"the quantum mechanical description of reality given by the wave function is not complete....*" (p. 126). Because the original article is quite technical, I shall present the clearer version described by Heinz Pagels, the underlying principles being the same.

We begin by measuring from a common reference point the positions q_1 and q_2, of two particles S_1 and S_2 which are moving with momenta p_1 and p_2. Although Heisenberg's uncertainty principle precludes measuring precisely the position and the momentum of either particle at the same time, we are permitted "to measure the *sum* of the momenta $p = p_1 + p_2$ and the *distance between* the two particles $q = q_1 - q_2$ without any uncertainty."[31] After interacting, one of the particles, S_2, radiates to London while the other, S_1, remains in New York. The principle of local causality (that the behavior of objects is determined by local causes, not by "superluminal" or "transtemporal" effects) implies that the two particles are sufficiently space–separated (too far apart for any light signal to reach them within the temporal limits) for S_1 to have any affect on S_2. But since the total momentum of the particles is conserved, we can deduce the momentum p_2 of the particle S_2 in London by subtracting the measured momentum p_1 of the particle S_1 in New York from the total known momentum: $p_2 = p - p_1$. Similarly, by measuring exactly the position q_1 of the particle S_1 (remembering that Heisenberg's principle does not preclude an exact *independent* measurement of position and momentum), we can calculate the position q_2 of the particle S_2 "by subtracting the known distance between the particles, $q_2 = q_1 - q$" (*ibid.*). Though measuring the position of the particle in New York will affect its momentum, this should not affect the momentum of the particle in London.

Thus both the position and momentum of the particle S_2 in London have been deduced without any uncertainty, contrary to Heisenberg's principle. Since according to quantum theory the position and momentum of the particle S_2 in London should be indeterminate, their exact deduction according to the EPR experiment led Einstein, Podolsky, and Rosen to conclude that quantum mechanical explanations are either incomplete, or that some mysterious affect was conveyed from particle S_1 in New York, at the moment it was measured, to particle S_2 in London, fixing the latter's position and momentum, in violation of the principle of local causality. Assuming that no one would reject local causality for a mysterious, instantaneous interconnection between space–like separated entities, the authors believed that they had proved the "incompleteness" of quantum mechanics. As Einstein says in his "Auto-biographical Notes:"

> But on one supposition we should, in my opinion, absolutely hold fast: the real factual situation of the system S_2 is independent of what is done with the system S_1, which is spatially separated from the former. . .(One can escape from this conclusion only by either assuming that the measurement of S_1 ((telepathically)) changes the real situation of S_2 or by denying independent real situations as such to things which are spatially separated from each other. Both alternatives appear to me entirely unacceptable.)[32]

Bohr's rebuttal was that the EPR experiment really begged the question since it assumed that because the position and momentum of the particle S_2 in London could be precisely *deduced,* this established these magnitudes as independent, *inherent* properties of the particle S_2, *apart from their being measured.* But this is exactly the point at issue. Bohr replied that while we might be able to deduce these magnitudes *before* measuring them, it is *only when they are measured* that their actual, albeit indeterminate, values are known. In other words, the inferred values are not the actual ones because the latter can only be established by measurement.

> From our point of view we now see that the wording of the. . .mentioned criterion of physical reality proposed by Einstein, Podolsky and Rosen contains an ambiguity as regards the meaning of the expression "without in any way disturbing a system." Of course there is in a case like that just considered no question of a mechanical disturbance of the system under investigation during the last critical stage of the measuring procedure. But even at this stage there is essentially the question of *an influence on the very conditions which define the possible types of predictions regarding the future behavior of the system.* Since these conditions constitute an inherent element of the description of any phenomenon to which the term "physical reality" can be properly attached, we see that the argumentation of the mentioned authors does not justify their conclusion that quantum–mechanical description is essentially incomplete (pp. 138–139).

While most physicists accepted Bohr's analysis, the controversy trailed some lingering doubts. In order to settle the issue once and for all, in 1964 a physicist at CERN named John S. Bell published a mathematical proof, known as Bell's theorem, that would enable physicists to determine experimentally whether quantum mechanics negated local causality.[33] Depending on how the results are interpreted, this could have tremendous implications. According to Gary Zukav:

> Bell's theorem...could affect profoundly our basic world view. Some physicists are convinced that it is the most important single work, perhaps, in the history of physics. One of the implications of Bell's theorem is that, at a deep and fundamental level, the "separate parts" of the universe are connected in an intimate and immediate way.[34]

The initial experiment devised by David Bohm to test Bell's theorem consisted of a two-particle system of zero spin so that when separated, each particle had a spin of opposite orientation: when one was oriented up the other was oriented down, when one was right the other was left. The spins were always equal and opposite. By using a magnetic field called a Stern-Gerlach device, one could orient the particles in a specific direction, up or down, right or left. Suppose that once the particles were separated and in flight one of them passed through the Stern-Gerlach device so that its orientation was changed. Would we expect this to affect the orientation of the second particle flying in the opposite direction? The answer is "no" if local causality holds. Bell's theorem mathematically prescribes a *limitation on the number of correlations* that can be experimentally detected between the spatially separated components of a two-particle system *if the principle of local causes is valid*. In other words, if the Bell *inequality* of correlations holds, then local causality or Einsteinian separability is confirmed; but if Bell's inequality is *violated* because there are *more correlations than expected*, then the EPR effect is supported and local causality denied.[35] In 1972 John Clauser and Stuart Freeman performed an experiment to determine whether the statistical correlations predicted by quantum mechanics *violated* the limitations specified by Bell's theorem or whether local causality was maintained, the results confirming the correlations predicted by quantum mechanics. More recently, an experiment by A. Aspect, P. Grangier, and G. Roger at the Optics Institute of the University of Paris "found that Bell's inequality was violated and resoundingly so (the discrepancy exceeded 13 standard deviations), thus upholding quantum mechanics."[36] According to physicists such as David Bohm,[37] Jack Sarfatti, and Henry Stapp, along with Gary Zukav,[38] these results necessitate a profound revolution in our conception of the nature of physical reality and its interactions. For physics is still based on the supposition that the uniform processes manifested in nature can be explained in terms of underlying particles (with wave properties) whose interactions depend upon physical impact or forces that act locally (recall the Grand Unified Theory described in the last chapter). However, according to the interpretation of the violation of Bell's theorem by the aforementioned physicists, the independence and separateness of physical events, along with local causality, are only apparent phenomena on the macroscopic and molecular level, obscuring a deeper interconnectedness and unity underlying physical reality. In Kantian terms, this intrinsic unity characterizes a reality in itself that appears to human beings as separate events localized in space–time, and interacting according to local causal influences. Moreover, this interpretation would confirm the insights of absolute idealists such as Hegel and F. H. Bradley, intuitionists such as Bergson, and Eastern mystics who claim that the diversity and separateness of the natural world is merely an apparent manifestation of a deeper, unifying reality more akin to spiritual or mental energy than to inert matter.

The above interpretation, however, is not the only possible one. Heinz Pagels, for example, attributes the unexpected quantum mechanical correlations *not* to some "instantaneous," "telepathic," or "superluminal" communication, behind the back, as it were, of space–time and local causality, but to mere chance correlations of two independent random sequences. Although the records of the orientations of the two particles represent two independent random sequences, the cross–correlation of the records indicate certain coincidences.

> We conclude that. . .Bell's experiment does not imply actual nonlocal influences. It does imply that one can instantaneously change the cross–correlation of two random sequences of events on other sides of the galaxy. But the cross–correlation of two sets of widely separated events is not a local object and the information it may contain cannot be used to violate the principle of local causality.[39]

The philosopher of science, Abner Shimony, also rejects the radical interpretation of Bell's theorem. But further discussion of this fascinating issue would carry us too far afield. What these divergent viewpoints illustrate is the dependence of the interpretation of the same experimental evidence on one's basic presuppositions and commitments.

This discussion of quantum mechanics also illustrates how different criteria of truth can influence one's acceptance and interpretation of experimental results. In general, while the *correspondence* criterion is particularly applicable to assertions in ordinary natural languages learned as we acquire our experience of the world, and the *pragmatic* criterion applies to hypothetical assertions whose test depends upon the possibility of actualizing the confirming situation, the *coherence* criterion applies especially to assertions concerning our most abstract theoretical frameworks. As I have continually emphasized, because the world as directly experienced and described in everyday language does not reveal its own raison d'être, we have had to create theories to represent the internal structures and extensive background conditions on which the foreground of experience depends. Initially, these theories consisted of analogical models copied from the operations of the world around us so that the processes could be imagined, but now that our theories have become so dependent upon exceedingly esoteric experimental data and highly abstract mathematical formalisms, the test of truth consists of internal consistency and congruence with the experimental evidence, namely, coherence.

I have also argued that experimentation has made the great difference in the advance of science because it provides additional necessary information or clues regarding the deeper structures and interactions of the universe on which the more superficial forms and organizations of matter depend. But these necessary experimental results would be unattainable and of little meaning if they were not the product of a framework of interpretation. As the last two decades in particle physics demonstrate, sometimes the theoretical developments outpace the experimental evidence, challenging experimentalists to design ingenious ways of confirming the predictions of the theories, while at other times the accumulation of experimental data awaits an Einstein, but the interaction of both is essential for the growth of science. Though experimental evidence is necessary for developing and testing theories, to be significant it must be intelligible in terms of some theoretical interpretation. A new discovery or experimental consequence that does not conform to our theoretical deductions and predictions indicates that something is amiss somewhere in the strands of our conceptual network, but not necessarily where. We then begin to question previous assumptions and critically examine the relevant concepts to resolve the difficulty. This process of theoretical adjustment and alignment is guided by the coherence of the interrelated factors. Although an unexpected experimental result can be sufficiently obdurate to signal an unsuspected feature of physical reality (such as Planck's quantum of action or the constant velocity of light), even this is possible only in terms of some interpretation.

Without the organization of astronomical observations within the two-sphere Ptolemaic system, the cumbersomeness, ad hocness, and discrepancies that disturbed Copernicus would not have been apparent. Similarly, the Michelson-Morley experiments with their startling outcome would not have been performed in the absence of the underlying presuppositions of the Newtonian framework. All our probing investigations of nature depend upon some background system of interpretation, but this does not prevent the recognition of discrepancies between theoretical implications and empirical data because there is enough constancy or uniformity in the interpretation to permit the recognition of conflicts, along with their resolution. When the discrepancy *is* global, it is either overlooked or ignored, as was true of the creationist view of the universe and man prior to Darwinism. This continual interplay between our theoretical interpretations and experimental results underlies the progress of science.

In conclusion, as the previous examples and discussions were meant to illustrate, when the question of truth is not merely one of the conformity of a statement within an accepted conceptual-linguistic framework to some previously classified state of affairs, fact, or determinate consequence, but is a question of the validity of the framework itself—what overall conceptual system one should adopt—one appeals especially to the coherence criterion. All the deepest questions of truth pertain to the validity of some conceptual framework as such, and how the various data are seen to be assimilated into that framework, whether it be a primitive mythology, a supernaturalistic religious system, the conceptual scheme of common sense and ordinary language, the geometrodynamic framework of Einstein's general theory of relativity, or the Grand Unified Theory of quantum mechanics. Each framework attempts to characterize the essential features of 'reality' according to its categorical scheme as determined by its methodology, but insofar as the seemingly inexhaustible complexity and fullness of reality overflows any conceptual system, none can claim absolute authority or finality. Nonetheless, these conceptual systems are the only intelligible way of representing what is true and real, since as human beings we cannot transcend the limits of the human condition in any other manner (unless psychical or mystical experiences turn out to have a greater validity than we in the West generally accord them). Though the methodology of modern science has been devised deliberately to subject frameworks to continuous rigorous testing to ensure the maximum objectivity or accommodation to the universe, we can never know what or how much has been omitted from any framework that would have to be included for a really true account of things. To paraphrase Wittgenstein, the possibility of describing the world in any one framework tells us less about reality than does the fact that it is capable of being represented in that way;[40] or, as C. I. Lewis asserted, our categories of interpretation "are divisions within the comprehensible in general, but not the shape of comprehensibility itself."[41]

Notes to Chapter VIII

1. C. I. Lewis, *Mind And The World Order* (New York: Dover Publishing Co., 1929), p. 259.
2. Paul Churchland, *Scientific Realism and the Plasticity of Mind* (Cambridge: Cambridge University Press, 1979), p. 2. Cf. Benjamin Lee Whorf, *Language, Thought, and Reality,* ed. by John Carroll (New York: MIT Press and John Wiley & Sons, 1956), pp. 57–64, 207–219.
3. Ludovico Geymonat, *Galileo Galilei,* trans. by Stillman Drake (New York: McGraw–Hill 1965), pp. 44–45.
4. Paul Feyerabend, *Against Method* (London: Verso Edition, 1975), p. 17.
5. Cf. Giorgio de Santillana, *The Crime of Galileo* (Chicago: University of Chicago Press, 1955), pp. 107–108.
6. Cf. Gerald Holton, *Thematic Origins Of Scientific Thought* (Cambridge: Harvard University Press, 1973), ch. 9.
7. Cf. Albert Einstein, "On the Electrodynamics of Moving Bodies," reprinted in *The Principle of Relativity,* H. A. Lorentz, A. Einstein, H. Minkowski, and H. Weyl, trans. by W. Perrett and G. B. Jeffery (New York: Dover Publishing Co., 1923), pp. 37–65. Also, Holton, *op. cit.,* ch. 5.
8. W. V. Quine, "Speaking of Objects," in *Ontological Relativity and Other Essays* (New York: Columbia University Press, 1969), pp. 16–17.
9. Cf. U. T. Place, "Is Consciousness a Brain Process?", reprinted in C. V. Borst (ed.), *The Mind–Brain Identity Theory* (London: Macmillan & Co., 1971), pp. 42–51.
10. Cf. E. E. C. Smart, "Sensations and Brain Processes," reprinted in C. V. Borst, *op. cit.,* pp. 52–66.
11. Cf. D. M. Armstrong, "The Nature of Mind," reprinted in C. V. Borst, *op. cit.,* pp. 67–79.
12. Cf. Paul Feyerabend, "Materialism and the Mind–Body Problem," reprinted in C. V. Borst, *op. cit.,* pp. 142–156.
13. Cf. Paul Churchland, *op. cit.,* p. 96. Also, *Matter and Consciousness* (Cambridge: MIT Press, 1984), pp. 56–61.
14. Cf. Chapter III.
15. Cf. D. M. Armstrong, *A Materialist Theory of Mind* (London: Routledge & Kegan Paul, 1968), p. 337.
16. Paul Feyerabend, "Materialism and the Mind–Body Problem," *op. cit.,* p. 152.
17. E. E. C. Smart, *op. cit.,* pp. 53–54. Brackets added.
18. Paul Churchland, *op. cit.,* p. 114.

19. Cf. Albert Einstein, "Autobiographical Notes," Paul Schilpp (ed.), *Albert Einstein: Philosopher-Scientist* (Evanston: Library of Living Philosophers, 1949), p. 85.
20. Rudolf Carnap, "Empiricism, Semantics, And Ontology," *Revue Internationale de Philosophie,* No. II, 1950, p. 22.
21. Albert Einstein, "On Physical Reality," *Franklin Institute Journal,* No. 221, 1936, pp. 349-350.
22. Werner Heisenberg, *Physics and Philosophy* (New York: Harper Torchbooks, 1958), p. 42.
23. A. Einstein, G. Podolsky, and H. Rosen, "Can Quantum-Mechanical Description of Physical Reality Be Considered Complete?", *Physical Review,* No. 47, 1935, pp. 777-780. Quoted from Stephen Toulmin (ed.), *Physical Reality* (New York: Harper Torchbooks, 1970), p. 123. Unless otherwise indicated, the following references are to this source.
24. Niels Bohr, "Can Quantum-Mechanical Description of Physical Reality Be Considered Complete?", *Physical Review,* No. 48, 1935, pp. 696-702. Quoted from Stephen Toulmin (ed.), *Physical Reality, op. cit.,* p. 132. Unless otherwise indicated, the following references to Bohr are to this source.
25. P. Laplace, *Introduction à la théorie analytique des probabilitiés, Oeuvres Complètes* (Paris, 1886), p. VI. Quoted from Milic Capek, *The Philosophical Impact of Contemporary Physics* (New York: D. Van Nostrand, 1961), p. 122. Italics added.
26. Cf. Norwood R. Hanson, "The Copenhagen Interpretation of Quantum Theory," *American Journal of Physics,* No. 27, 1959, pp. 1-15. Reprinted in Stephen Toulmin (ed.), *Physical Reality, op. cit.,* p. 146.
27. Niels Bohr, *Essays 1958-1962 on Atomic Physics and Human Knowledge* (New York: Vintage Books, 1963), p. 4.
28. Cf. John von Neumann, *The Mathematical Foundations of Quantum Mechanics,* trans. by Robert T. Beyer (Princeton: Princeton University Press, 1955).
29. Werner Heisenberg, *op. cit.,* p. 48.
30. Richard Rorty, *Philosophy and the Mirror of Nature* (Princeton: Princeton University Press, 1979), p. 178. Brackets added.
31. Heinz R. Pagels, *The Cosmic Code* (New York: Simon & Schuster, 1982), pp. 163-164.
32. Albert Einstein, "Autobiographical Notes," *op. cit.,* p. 85.
33. Cf. John S. Bell, "On the Einstein Podolsky Rosen Paradox," *Physics,* No. 1, 1946, p. 195ff.
34. Gary Zukav, *The Dancing Wu Li Masters* (New York: William Morrow & Co., 1979), p. 298.
35. Cf. Bernard d'Espagnat, "The Quantum Theory And Reality," *Scientific American,* November, 1979, pp. 158-181.
36. "Bell's Theorem: Still Not Ringing True," *Scientific News,* August 22, 1981, p. 117.
37. Cf. David Bohm, *Causality and Chance in Modern Physics* (New York: Harper Torchbooks, 1957), ch. IV, and *Wholeness And The Implicate Order* (London: Routledge & Kegan Paul, 1980), ch. IV. Also, E J.

Belinfante, *A Survey of Hidden-Variables Theory* (New York: Pergamon Press, 1973).

38. Cf. Gary Zukav, *op. cit.,* pp. 309–331.
39. Heinz R. Pagels, *op. cit.,* p. 176.
40. Cf. Ludwig Wittgenstein, *Tractatus Logico-Philosophicus,* trans. by D. F Pears and B. F McGuinness (London: Routledge & Kegan Paul, 1961), 6.342.
41. C. I. Lewis, *op. cit.,* p. 237.

CHAPTER IX

CONTEXTUAL REALISM

IN this concluding chapter I shall attempt to describe and justify a conception of 'reality' as implied by our previous discussion and present attainment of knowledge. Although in the past it was presupposed that in any limited time frame, such as the brief life span of an individual, it was possible to realize a timeless perspective or absolute view of 'reality,' the acute sense of historical change brought about by the theory of evolution, as well as the rapid succession of political, social, and intellectual developments churned up since the middle of the nineteenth century, have made us aware how fragile and transitory are even the most seemingly secure institutions and beliefs.

Just as all scientific knowledge appears to be conditioned by the method and context of investigation, so anyone attempting to sketch a 'meta-physical' view must be aware how difficult it is to see beyond the limited presuppositions of his own epoch. As I am using the term, "meta-physical view" designates the overall tentative conception of reality implied by our current knowlege of the universe, in contrast to the traditional term 'metaphysics' which, since the seventeenth century, has come to stand for an absolute system of reality based on some privileged mode of knowing, such as Descartes' "cogito ergo sum" or Bergson's "sympathetic intuition" of reality.[1] My use of the term 'meta–physical' corresponds more to the original meaning, acquired accidentally due to the editing of Aristotle's works, since Aristotle's treatise on "first principles" and "being *qua* being," now known as the *Metaphysics,* obtained its name allegedly by an ancient editor who placed it after (*meta*) physics.

How different the philosophical perspectives of Aristotle, Hume, and Kant would have been had they had access to the knowledge of today; similarly, the views of any contemporary thinker must necessarily rest upon unquestioned principles which reflect the presuppositions and limitations of our own era. Thus any pretensions to a *philosophia perennia* would seem to be rebuffed by history, although just as any age will have its savants and poets, so it has its philosophers who will attempt to mark, however imprecisely, the impression of 'reality' on experience. In the eloquent words of F H. Bradley:

> And so, when poetry, art, and religion have ceased wholly to interest, or when they show no longer any tendency to struggle with ultimate problems and to come to an understanding with them; when the sense of mystery and enchantment no longer draws the mind to wander aimlessly and to love it knows not what; when, in short, twilight has no charm—then metaphysics will be worthless.[2]

The 'meta–physical' view to be presented here is that of contextual realism,[3] an overall perspective similar to the "levels hypothesis" defended at one time by such scientists or philosophers as David Bohm,[4] Mario Bunge,[5] and John Cowperthwaite Graves.[6] As the subsequent discussion will indicate, the term "contextual realism" is preferred, in that it more graphically depicts the intended characterization of physical reality as *an inexhaustible matrix of relatively autonomous contexts,* while following as a natural successor to such previous epistemological (and ontological) positions as "naive realism," "phenomenalism," "Kantianism," and "radical critical realism." For reasons that I hope will be convincing, I think the conception of 'reality' implied by twentieth century scientific developments conforms more to the meaning of "contextual realism" than to any of the previously mentioned epistemological positions.

According to the meta–physical framework of contextual realism, reality consists of seemingly inexhaustible levels of semi-autonomous (or 'real') contexts exhibiting a myriad of forms, properties, structures, and processes. Analogous to the innumerable cellular structures disclosed at successive levels by various staining techniques as seen under different optical resolutions through a microscope, the world resolves into endless matrices of relatively stable contexts exhibiting phenomena subject to varying descriptive predicates and explanatory principles. Although these multifarious contexts with their various structures are not entirely discontinuous (otherwise our knowledge of them would be much more difficult than it already is), neither are they merely successive dimensions of essentially the same complexes of elements, as was assumed in the past. Though manifesting some analogous relations and properties, still, the forms and processes of the macroscopic world are not qualitatively similar to those of the atomic–molecular domain, and the conjugate properties of the latter, according to quantum mechanics, are not repeated on the subatomic level. Moreover, as one moves outward to cosmic, as opposed to inner atomic dimensions, one finds that the structural relations between space and time, force fields and mass, or gravitational fields and the space–time continuum become radically altered, as described in the general theory of relativity.

Thus a fundamental assumption of all past inquiry would seem to be refuted by recent developments in science; namely, the presupposition that any cross section of the universe will reveal a similar ontological grain or, less metaphorically, that inquiry at any dimensional level of the universe will disclose the same basic elements, underyling structures, and explanatory principles. This, in turn, implies the denial of two additional presuppositions of the past: (1) that the diversity of phenomena on the macroscopic level can be *fully* explained in terms of a finite set of unalterable, eternal elements and interactions discovered at a deeper level of investigation, and (2) that the laws and causal principles applicable in any one context of inquiry can be extrapolated *indefinitely* to all other levels of phenomena. If these assumptions are false, no final or absolute knowledge of the universe is possible because such certainty would depend upon knowing the universe in all of its diverse contexts or dimensions. Successful as we have been in exploring deeper and more extensive domains of the physical world, the total complexity and extensity of reality would seem to be beyond our finite grasp. As Karl Popper states:

> We must come to terms with the fact that we live in a world in which almost everything which is very important is left essentially unex-

plained. We try our best to give explanations and we penetrate deeper
and deeper into really incredible secrets of the world with the help of
the method of conjectural explanation. Still, we should always be
aware that this is only in a sense scratching the surface, and that
ultimately everything is left unexplained, especially everything in
connection with existence.[7]

Stated differently, the predominant assumption underlying the
modern picture of the universe and quest for knowledge (though, as
all our leading ideas, it had its origin in Classical Greece) has been
that the world is formed from a limited number of fundamental,
immutable, eternal elements. In some unfathomed manner, the interaction of these basic particles produced all the various domains
(macroscopic, subatomic, and cosmic) and manifestations (qualitative as well as quantitative, organic as well as inorganic, mental as
well as material) of the universe. According to this assumption,
ultimate knowledge consists of discovering the fundamental elements and universal forces of nature, described in terms of physical
properties and interactions expressed as equations within the theories, in contrast to the sensory qualities, perceptual forms, and
contingent events in the world as immediately experienced. Thus
'reality' consists of "things in themselves" as opposed to the world as
it appears to man, and if we possess any actual knowledge at all, it is
because our theories correspond to this reality, as Einstein claimed.

But Einstein also disclosed one of the two basic restrictions on
knowledge divulged by twentieth centry physics that undermined
these traditional assumptions, "limiting conditions." Einstein found
that while certain concepts appear to have an invariant meaning and
universal application independent of any conditions or contexts,
they actually break down when extended beyond certain limits.
Thus our concepts of temporal and spatial dimensions that are
invariant within their own "proper" coordinate systems acquire
different magnitudes for observers moving relative to them; similarly, events which are simultaneous for one coordinate system are
successive for systems moving relative to it. Only under the "limiting conditions" where the velocities and distances are insignificant
compared to the velocity of light (as they are for all practical
purposes in our daily experiences), can one ignore relativistic effects and use the concepts of space and time in their Newtonian
sense, along with the more familiar Galilean transformation equations. For example, gazing at the sky on a clear dark night, one is
deceived if he believes he is seeing each of the stars instantaneously

as they are now, rather than reacting to light signals which, because of their finite velocities, began at different space–time intervals remote in the past.

A similar situation pertains to quantum mechanics. Under the "limiting conditions" where the magnitudes of physical interactions are large compared to Planck's quantum of action, Heisenberg's principles of indeterminacy with their inherent probabilities do not take effect. Thus classical mechanics can predict precisely the trajectories of bullets, missiles, and space ships because their dimensions do not preclude measuring simultaneously and exactly such conjugate magnitudes as position and momentum, or energy and time. It is only in subatomic experiments such as radiation, where the physical dimensions of energy, mass, frequency, and wavelength are proportional to Planck's quantum of action, that the anomalies of quantum mechanics begin to appear. As Bohr stated:

> Just as the relativity theory has taught us that the convenience of distinguishing sharply between space and time rests solely on the smallness of the velocities ordinarily met with compared to the velocity of light, we learn from the quantum theory that the appropriateness of our usual space–time description depends entirely upon the small value of the quantum of action as compared to the actions involved in ordinary sense perceptions.[8]

The second encounter in twentieth century physics that imposed a radical restriction on knowledge, manifested in all aspects of quantum mechanics, has been brought out by Bohr in his "principle of complementarity." According to Bohr's principle, rather than quantal processes consisting of objective states with inherent properties that are disclosed under suitable experimental conditions, so that physical theories are true if they correspond to these antecedent realities, the vector states at the subatomic level consist of superpositions of properties with merely a probable value, whose determination depends upon how the experiment is prepared and what measuring apparatus is used. In this view, physical reality is not preformed or predetermined, in the sense that all the variables necessary for an exhaustive objective knowledge *could* exist at any one time: instead, they have an indeterminate, reciprocal status so that the degree to which one is ascertained limits the degree to which its conjugate can be manifested. Which of these exclusive but complementary components of the quantum object materialize is *partially* dependent upon how the scientific experiment is designed, so that the scientist is not merely a spectator, but an active participant in the actualization of nature in his investigations—as Dewey claimed as well. In Bohr's words, "the new situation in physics has so forcibly reminded us of the old truth that we are both onlookers and actors in the drama of existence" (*ibid.,* p. 119).

Though the discovery of reciprocally limiting but complementary states and properties had no place in classical physics, one can find considerable evidence for it in everyday experience, once one is alerted to the possibility. For example, now that the world's "micro–economic system" has become so interdependent and complex, economists are forced to admit that such conjugate economic conditions as higher growth rate and higher inflation, lower inflation and higher unemployment, or tight monetary policy and slow growth rate are so interdependent that whatever action is taken to deal with one condition adversely affects the other. In addition, the fact that so many well–intentioned government programs either do not succeed or lead to unexpected adverse effects, undoubtedly attests to the complex, contingent interdependence of the factors involved.

Another familiar area in which Bohr's view would seem to be clearly illustrated is that of ordinary perception. Just as Bohr emphasized that precise quantal states and their values do not preexist in the classical sense, but are a joint product of the interaction between the quantum object and the apparatus, with certain experimental preparations and results precluding others, so the manifestation of the world to human beings depends upon the interaction of the human sentient organism with the physical environment. Depending upon whether they stimulate optical nerve endings or receptor nerve cells in the surface of the skin, electromagnetic vibrations are experienced as colors or as tactual sensations of heat, pressure, or pain. Moreover, there is some evidence that quantum mechanics supports the Kantian view that space–time and causal determinism are not the fundamental pre–conditions of natural occurrences, but features of experience relative to the level of investigation: that is, that physical processes occurring in space–time coordinate systems according to determinate antecedent causes might actually originate in subatomic processes that lie outside these seemingly universal and necessary conditions of experience or phenomena. As the physicist Richard Schlegel indicated in the concluding dialogue of his book:

"Our world view is indeed changed by the strong evidence that quantum physics has given us for a nondetermined, statistical element in physical processes. But there is an even more profound philosophical change that comes with our new knowledge of the microdomain. The world is formed, in the detail of its individual events, only in the act of the appearance of those events. The classical physics concept is of a universe of particles and processes among which we move and live: they are *there* before, during, and after the time we in any way interact with them. But now we know that at the level at which Planck's constant is important—and how appropriately it is called the constant of action—the world that we describe takes its pattern of event and process as it is in the act of being observed, or, in general, in an act of coming into some one state as it interacts with some other part of the world."[9]

Paradoxically, the disclosure of the two limitations on knowledge described above were directly related to the discovery of two universal physical constants: the invariant velocity of light in relativity theory and Planck's constant of action in quantum mechanics. So even here one can detect an element of complementarity: the discovery of certain constants in nature complementing the disclosure of the limitation of concepts previously believed to have an unrestricted application.

Thus developments in twentieth century science have decidedly undermined man's previous confidence in the possibility of attaining absolute or final knowledge—issuing in what could justifiably be called the age of uncertainty. Neurologists and philosophers have reached an "impasse" in their attempts to explain or reduce conscious experiences to neurological processes. Though quantum mechanics is the most powerful theory yet developed in physics, quantum phenomena are themselves described as "inscrutable" and "unpicturable," comprised of both wave and particle characteristics, and existing in an indeterminate probabilistic state of superpositions. Furthermore, all attempts to discover the "basic elements" or "building blocks" of physical reality have led to the annihilation of the purported fundamental entities with the creation of new particles or energy, as described by the Feynman diagrams or S–matrix theory. As Sir Karl says, in an exact expression of contextual realism:

For a long time, essentialism had been identified by all parties, including its positivist opponents, with the view that the task of science (and of philosophy) was to reveal the ultimate hidden reality behind the appearances. It has now turned out that although there are such hidden realities, none of them is ultimate; although some are on a deeper level than others.[10]

We now realize that the traditional attempt to define this ultimate reality in terms of the scientific–philosophical distinction between primary and secondary qualities—the former being "intrinsic properties" in the insensible particles existing independently of our experience of them, and the latter merely "subjective additions" of the mind—is as arbitrary as it has always appeared. Both primary and secondary qualities are 'real,' relative to certain conditions; they are both inherent (though variable), intersubjectively confirmable properites of macroscopic objects as they exist *within the context of ordinary experience.* But alter the conditions sufficiently, either by modifying the organism and/or the physical stimuli, and the manifest properties of things change, whether they be the sensory (secondary) qualities or the physical (primary) qualities. For example, if we were sensitive to ultraviolet light the visual qualities of the ordinary world would be radically different, while if we possessed "microscopical vision" the primary qualities of macroscopic objects, such as their solidity, shapes, and discrete positions, would vanish. Although one can *theoretically* abstract the secondary from the primary qualities, the attempt to *actually* peel off the former, leaving ordinary objects with the latter, is hardly tenable. To extraterrestrial creatures, if there be such, the world would undoubtedly appear quite different than it does to us, as regards both primary and secondary qualities.

Accordingly, in contrast to the traditional conception of the world composed of a finite number of fundamental particles and structures, recent developments in science suggest that the universe consists of innumerable contexts of endless varieties of entities and processes which 'really exist' within certain conditions. If, as it seems, every domain of discoverable phenomena depends upon a more fundamental matrix of background conditions and internal substructures, in an endless series of broader and deeper contexts, then no one context, other than the absolute totality of realty, can be singled out as final. As David Bohm expresses this,

we assume that the world as a whole is objectively real, and that, as far as we know, it has a [somewhat] precisely describable and analyzable structure of unlimited complexity. This structure must be understood with the aid of a series of progressively more fundamental, more extensive, and more accurate concepts, which series will furnish, so to speak, a better and better set of views of the infinite structure of objective reality. We should, however, never expect to obtain a complete theory of this structure, because there are almost certainly more elements in it than we can possibly be aware of at any particular stage of scientific development. . . .The point of view described above evidently implies that no theory, or feature of any theory, should ever be regarded as absolute and final.[11] (Brackets added.)

This is not to claim that such a contextual view is the only possible or plausible interpretation of the current stage of scientific knowledge, or that all scientists would agree with the above characterization. Graves, for example, while also adopting the hypothesis of levels, believes that the "geometrodynamic framework" of the general theory of relativity is, in principle, capable of accounting for all the various strata of the universe.[12] However, it seems to me unlikely that we could explain all the marvelous diversity and extraordinary forms in the world, such as the beautiful variety of flora and fauna merely in terms of a basic space–time framework. Also, as we saw in the last chapter, there are scientists who believe that the recent developments in particle physics leading to the Grand Unified Theory indicate that physicists are on the threshold of a complete explanation of the nature of the atom and of the structure and origin of the universe, as described in the Big Bang and "inflationary universe" theories.

Again, however, given the limitations of human experience in comparison with the tremendous complexity and seemingly inexhaustible dimensions of the world, it is difficult to believe that we could extrapolate from a few apparent constants to anything like a complete knowledge of the universe. Obviously, such an assessment itself should only be made on a contingent, undogmatic basis, for no one knows what surprising discoveries lie ahead which might lead to more conclusive knowledge, or which at least would suggest a different picture of reality than that of contextual realism. But as the physicist Freeman Dyson stated in answer to a question in a recent interview:

And what of man's immemorial quest for a single, unified set of laws by which to subjugate the cosmos? "If it *were* like that," Dyson sighs, contented, "then the creator lacked imagination. I would find it disappointing if He had made it all so that it could be understood with one equation. My feeling is that physics is inexhaustible. The deeper we dig, the more layers we'll discover, so that it goes on essentially in all directions forever."[13]

This is precisely my point of view, although I would not restrict it to physics. It was also the view of Isidor Rabi, Nobel Prize winner, and for many years the distinguished Chairman of the Physics Department at Columbia University: "I don't think that physics will ever come to an end. I think the novelty of nature is such that its variety will be infinite — not just in changing forms but in the profundity of insight and the newness of ideas. . . .[14] That inquiry should be an unending search, rather than a final attainment, has its own lure!

The claim that all knowledge is relative to limited conditions and contexts, albeit some more basic and pervasive than others, should not be interpreted to mean that no confirmable truths or reliable knowledge are attainable. The very fact that certain contexts are relatively autonomous (or real) makes it possible to acquire a definite, if not exhaustive, knowledge of nature as manifested within those conditions. In fact, the more restricted the context and limited the claim, the more certain knowledge becomes. To insure the maximum truth value of a statement, we have only to reduce to a minimum the factual claim made by the statement; thus, the assertion that a particular object *seems* to feel hot is more certain than the statement that the object *feels* hot, or that it *is* hot, or that the temperature of the object is *due* to the mean kinetic energy of the molecules making up the object.

This restriction of truth claims explains the force of Descartes' "*cogito*" argument, Wittgenstein's "tautologies," and Moore's "defence of common sense." In each case, the claim to certainty is proportional to the risk taken: in Descartes' case, to the inference that being conscious implies his existence with greater certainty than anything else he could know, to Wittgenstein's definition of tautologies (such as "a rose is a rose") as being true under all truth conditions, to Moore's claim that we know with certainty various ordinary propositions describing certain common features of the world, so that to deny the truth of these propositions would be to deny the very aspects of the world necessary for our existence. Thus the following paradox: in general, the more certain our knowledge, the more trivial it becomes, while the more significant it becomes, the less certain it is.

As pointed out in Chapter VI, our most assured factual knowledge consists of assertions designating the world as it is 'directly' experienced—hence the seductive appeal of naive realism, common sense beliefs, and ordinary language. What could be more certain than the facts that grass is green, that the earth is solid, that fire is hot, that each of us was born in the past? Such facts announce themselves like emblematic signs hanging from British pubs. Unfortunately, however, the certainty of such statements belie their significance. What does one know when he knows that grass is green (how chlorophyll is synthesized?), that the earth is solid (the geological composition of the earth?), that fire is hot (the nature of radiant energy?), or that the present was preceded by past events (the nature of time?)? The facts described by such statements do not close inquiry, but provoke it.

As described earlier, the manner in which our central nervous system exhibits the world to us is accomplished so covertly that we are largely ignorant of how it happens. Though we conjecture that our brains function like television receivers transforming electromagnetic stimuli into images of the world on the screens of our consciousness, or that the cells of the brain act like a Polaroid film fixing on our consciousness pictures of the world due to the patterns of light filtered through our eyes, or that our perceptual experience is a holographic projection caused by the interference of light waves striking the retinae, such analogies are not very exact and tell us very little about the actual process. Although under normal conditions we encounter a world that is outside us in space, that is objectively or publically observable, and that exerts causal effects on our bodies so that our existence is dependent upon the positive influences of these effects—all of which attest to the independent existence of the world—somewhat analogous features of experience are reproduced in dreams (as Descartes argued) when the cause lies within us (or within our brains), rather than in the external world. Thus, for all its apparent independent reality, the perceived world is not unlike a dream, as poets and mystics have declared throughout history.

This enigma is further enhanced by the fact that while the macroscopic world appears initially to be a self-contained, self-sufficient reality, all attempts to understand and explain natural occurrences have led to the belief in transcendent causes, whether as primitive daemons, the Will of God, Newton's forces and indestructible atoms, Kant's things in themselves, or the latest particles, such as quarks, postulated by the physicist. In addition, the conviction that the world, as disclosed to us under normal conditions, is really just a kind of surface appearance or manifestation of a much more extensive reality was confirmed by the development of the telescope and microscope which provided additional 'direct' experience of further domains of phenomena normally hidden from our perceptions, while the use of spectroscopes, Wilson cloud chambers, Geiger counters, and computerized graphic simulations provide *indirect* evidence of even vaster dimensions of reality.

But just as our predecessors were unaware of the existence of molecules, mesons, and neurons, how much more lies hidden in the depths of reality than we can ever know at any one time? Moreover, and this needs to be particularly emphasized, it should not be taken for granted that only the type of experimental investigation devised by modern science is capable of discovering new entities or of disclosing hidden dimensions of nature: *there are undoubtedly ways, unknown to us today but which will be stumbled upon in the future, of disclosing undreamed secret recesses in nature that will be just as effective as the experimental method of science for discovering new entities, structures, properties, and processes.*

Accordingly, the transparent truth of ordinary statements and the certainty of common sense beliefs within the context of normal, everyday experience should not lull one (as it did ordinary language philosphers) to the belief that this context is unconditioned or self–sufficient. Even though scientific statements describing postulated entities and theoretical domains, as disclosed under experimental conditions usually cannot be *directly* confirmed, their explanatory significance and predictive testability more than compensate for their lesser certainty in relation to their mode of access and verification. As is now generally conceded, the program of the early positivists restricting the reference of scientific statements to either the phenomenalistic content of experience, as Mach originally advocated, or to the ordinary physical world of macroscopic objects, instruments, and pointer readings, as Schlick and Carnap later decreed, on the grounds that these epistemic foundations were the only certain or secure ones, has been discredited by developments within science itself—it being futile to insist on the self–contained, self–sufficiency of the macroscopic world when it cannot account for its own existence. Even the uncertainty as to the exact nature of subatomic structures and processes, and the restriction of matrix mechanics to observable magnitudes, did not lead quantum physicists to deny that the quantum object exists in *some* sense—

that the interaction between the quantum object and apparatus was essential!

Ironically, Dewey accused the classical tradition of ancient Greece, as well as classical modern science (along with the Western religious tradition), of being preoccupied with the "quest for certainty," despite the daring originality and soaring achievements of both traditions. But no schools have been more obsessed with achieving certainty in knowledge than twentieth century positivism and ordinary language philosophy. The positivists attempted to attain this certainty by grounding all knowledge on an indubitable foundation of sense data reinforced by a logically tight artificial language devoid of ambiguity and impervious to metaphysical assertions, while ordinary language philosophers invoked "common sense" and the discourse of "the plain man" as their securities against speculative claims. Though both schools have had beneficial effects in calling attention the crucial role of language in philosophizing and in making contemporary philosophy more rigorous, these positive results were greatly overshadowed by their shortsighted insistence that the "paradoxes," "impasses," and "crises" of knowledge confronting us on all sides are not symptoms of the limitations of knowledge, but are "conceptual muddles" or "pseudo–problems" arising from the misuse of language attributable to excessive speculative or metaphysical zeal. Thus the impression was conveyed to several generations of philosophers that all problems would eventually be solved if only we had sufficient confidence in science and did not raise any spurious philosophical questions—philosophy, or at least metaphysics, being made the scapegoat for unanswerable, contentious problems. A less imaginative, more *trivializing* conception of the problems of knowledge and of the role philosophy could hardly be imagined.

This view of the positivists is clearly advanced in the following statement by Moritz Schlick, founder and leader of the Vienna Circle:

> ... I am convinced that we now find ourselves at an altogether decisive turning point in philosophy, and that we are objectively justified in considering that an end has come to the fruitless conflict of systems. We are already at the present time, in my opinion, in possession of methods which make every such conflict in principle unnecessary....These methods...have their origin in logic. Leibniz dimly saw their beginning. Bertrand Russell and Gottlob Frege have

opened up important stretches in the last decades, but Ludwig Wittgenstein. . .is the first to have pushed forward to the decisive turning point. . . .[15]

Having pointed out the new logical methods that were contributing to the "turning point in philosophy," Schlick expressed his confidence that there were no questions that were "in principle unanswerable."

> This. . .has consequences of the very greatest importance. Above all, it enables us to dispose of the traditional problems of "the theory of knowledge." Investigations concerning the human "capacity for knowledge," insofar as they do not become part of psychology, are replaced by considerations regarding the nature of expression, of representation, i.e., concerning every possible "language" in the most general sense of the term. Questions regarding the "validity and limits of knowledge" disappear. Everything is knowable which can be expressed, and this is the total subject matter concerning which meaningful questions can be raised. *There are consequently no questions which are in principle unanswerable, no problems which are in principle insoluble. What have been considered such up to now are not genuine questions, but meaningless sequences of words* (*ibid.,* pp. 55–56; italics added).

In light of the continuing controversy over the paradoxes in quantum mechanics and whether the theory is complete or incomplete, as well as the current "impasse" in the mind–body problem, such confidence seems overwrought. Moreover, it has not just been philosophers (such as Hume and Kant, Russell and Popper) who have raised questions as to the "validity and limits of knowledge," but physicists such as Einstein, Bohr, Heisenberg, and Bohm, as well as neurophysiologists such as Penfield, Eccles, and Sperry.

Returning to the discussion of contextual realism, while the world as experienced appears to be self-contained and self-sufficient, cataclysmic changes in nature, as well as elaborate experiments, exhibit the contingent or conditional existence of all physical phenomena relative to more pervasive background (environmental) conditions and internal (atomic–molecular) substructures. There is hardly anything in the universe as we know it that cannot be transformed into something different, no substance or element that cannot be decomposed or destroyed, by sufficiently modifying the external conditions and/or its internal composition. The geological forms found on the earth could not exist in the thermo-nuclear environment of the sun, but even the composition of the sun is different from that of the original, incredibly condensed state of the universe postulated by the Big Bang theory. Though the geological surfaces of the earth and the moon are somewhat similar, the temperature and atmospheric conditions (as they now exist) on the latter do not support the existence of life as found on the earth.

As regards the earth itself, five to eight billion years ago it was nothing but an accretion of gas and dust in the nebula surrounding the sun; but as this amorphous state cooled and condensed forming our planet with its seas and enveloping atmosphere, the conditions arose which made possible, and over an extended period of time probable, the "creation," "formation," or "emergence" of the simplest forms of organic structures. Then, from these original single-celled organisms, there developed multiple-celled creatures which produced the incredible diversity of flora and fauna, first in the sea and then growing, crawling, flying, and walking on the earth. How these various organic forms, with all their intricate structures and functions (the extraordinary variety of flowers, of fish, and of insects), evolved from the original conditions and simple biomolecular structures seems to be as ultimately unexplainable as the origin of all the other forms and qualities of phenomena that have arisen under various natural conditions. For though scientists can describe, and even partially explain, how certain kinds of compounds originate and physical transformations occur, they are usually unable to explain why such compounds have the properties they do, and why the transformations occur as they do: why $NaCl$ has the macrosopic qualities of salt and HCl of hydrochloric acid; how the ordinary macroscopic world comes to appear to us as it does; or how nerve firings eventuate in conscious experiences.

One could present numerous examples from physical and organic chemistry, metallurgy, molecular biology, and other disciplines to further illustrate the conditional status of phenomena. But our own normal biological functioning is perhaps the most immediate example of an existence which seems to be self-sufficient but which, in fact, is dependent upon a complex, delicate homeostatic relation with the external environment in terms of air, temperature, pressure, humidity, nutrients, and radiation. Still, select any stable object in the environment and, by changing the conditions sufficiently, one can modify, transform, or destroy it.

Thus while our experience of the world is characterized by the illusion of discrete independence, scientific investigations of physical phenomena disclose the complex, interrelated conditions within which natural states and processes occur. Undoubtedly for survival reasons, our nervous system has evolved in such a way as to form organs that abstract from this multiplicity certain determinate shapes, structures, and properties, representing them as independent and self–existent. As many others have insisted, reality is like an immense tapestry with its surface design exposed to our direct awareness, while all the intricate network of underlying connections are hidden from view.

But if reality did not manifest such seemingly independent patterns and forms, despite their substructural interconnectedness, there would be nothing to support our usual linguistic conventions. For this reason, *"naive"* or *"direct* realism" seems such a natural and valid point of view. *Within the context* of ordinary experience it does appear that the world presents itself as it is, as if our experience does not affect or transform its character, but merely conveys its real features to us as they are. However, scientific revolutions (such as those of Copernicus and Einstein) and empirical discoveries (as in atomic physics and neurophysiology) have forced us to discard the naive, common sense point of view, natural as it may be. For this reason some philosophers and scientists have turned to a radical form of *"critical* realism,"[16] identifying the 'real world' with the conception of reality conveyed by the physical sciences, and relegating the ordinary world of experience to the status of "appearance," "subjectivity," or "mind–dependence." It has even been proposed that the "manifest image" of the world of ordinary experience will some day be assimilated into the purely physicalistic framework of science, so that all references to phenomenal qualities and subjective experiences will "disappear."[17]

So while "naive" or "common sense realists" maintain that the ordinary macroscopic world is the real world, and that the scientific conception of physical reality must be dubious or fictitious (despite its undoubted explanatory value), "radical critical realists" accept the scientific representation of the world as the real one, treating the ordinary macroscopic world as merely a subjective phenomenon (despite its unquestioned existential and epistemic primacy). However, neither position, it seems to me, does justice to the complexity of the world as we understand it today. Given the overwhelming evidence in contemporary science for some kind of

infrastructure underlying ordinary phenomena, and given the transforming theoretical and practical impact on our lives due to the
scientific exploration of such domains—which could hardly be
accounted for if they were merely "fictional"—the rejection of any
reality beyond the macroscopic world seems not only naive, but
incredible! Conversely, the position that the ordinary macroscopic
world, which is the beginning and end point of all scientific research, has merely a subjective status without any objective reality,
occurring in our individual heads, as it were, does not do justice to
our existential sense of *existing within* and *being dependent upon*
this ordinary world—especially as there is no explanation of how
our central nervous system performs this extraordinary, duplicitous
inversion.

Thus the thesis that both contexts have a certain degree of
autonomy or reality relative to certain conditions, fits the metaphysical theory of contextual realism which takes into account both
the conditional or "contextual" existence of all reality, as well as its
autonomous or "realistic" status within such conditions. The ordinary context of experience that for millenia seemed to be the
absolute face of reality—the domain that Aristotle considered fundamental in opposition to the Atomists—is now relegated to the
macroscopic level, despite its self-contained, autonomous appearance. Obscure as the origins of some phenomena, such as
diseases, births, deaths, and natural catastrophies have been, most
occurrences in nature initially appear to be the outcome of *observable* conditions and causes: trees grow, fire devours trees, water
quenches fire, freezing temperature crystalizes water, the sun dis-
solves ice, and so on. Normally, there seem to be no gaps or discontinuities in the eternal recurrences of nature, only temporary states
or continuous transitions. Objects localized in space have their own
inherent potentialities and temporal destinies marked by certain
fates. And though the actualization of things is seen to depend upon
various conditions, these conditions appear to be inherent in the
macroscopic domain itself: the substantial natures, essential properties, and normal functioning of phenomena coinciding with their
appearances in this domain.

The development of modern science shattered this composed
image of reality. A closer, critical, more exact scrutiny of natural
occurrences revealed their dependence on deeper causes and structures. Investigations of vision and optics indicated that the qualities
of the perceived world depend upon the nature of our sense organs,

and therefore—however counterintuitive this might appear—cannot exist apart from us. Roemer's proof in 1675 of the finite velocity of light also implied that what we sense is based on stimulation of our peripheral sense organs, so that our perceptions occur in some way in us. Then, with the construction of telescopes and microscopes and later of even more sophisticated instruments, we were compelled to acknowledge much vaster domains of physical reality than recorded by our limited sense receptors. Our conception of nature was expanded to include the cosmic and microscopic domains, along with the middle (or macro) level of existence.

With continuous refinements in techniques of investigation and advances in the technology of research equipment, the microscopic world itself came to be differentiated into five domains or levels: the molecular, atomic, nuclear, subnuclear (consisting of the hadrons and electroweak and strong forces), and now the level of quarks. Although the organization of matter and the occurrence of events on the macroscopic level appear to be self-contained and self-explanatory, we have been compelled to acknowledge they are not. The human organism functions as an autonomous unity, but is nonetheless composed of systems (digestive, nervous, reproductive, etc.), organs, tissues, cell assemblies, and cells. The cell, the basic unit of living organisms, itself consists of more elementary components, the nucleus containing the chromosomes, the cytoplasm, the cell wall, and so on. Similarly, all the diverse substances in the macroscopic world have inner structures, a molecular structure, on which their observable characteristics and causal powers depend. These molecular structures are the smallest units capable of retaining the chemical properties of the diverse substances and constitute their internal natures.

It is one of the strangest features of the world that while substances such as water, salt, wood, and diamonds exist on the macroscopic level as self-sufficient entities characterized by observable qualities and causal properties, their seemingly autonomous states are ultimately dependent upon, and therefore explainable in terms of, intrinsic structures. While some of these 'objective' characteristics, such as the sensory qualities of color, taste, smell, and tactual sensations, occur only within the context of an interaction with an organism possessing sense receptors such as ours, other properties depend upon the inherent molecular structures themselves. The fluidity, transparency, temperature, and solvent properties of water, for example, are due to its molecular structure and the strength of its chemical bonds. External conditions independent of us, such as temperature and energy, can affect these bonds transforming the liquid substance to ice or vapor. Moreover, it is in terms of these molecular structures that we account for the chemical reactions of substances. With this knowledge we have been able to create new substances, like gasolines, plastics, synthetic fibers, and various vaccines and drugs.

Although our ordinary visual acuity cannot disclose these infrastructures, they can be discovered by physical and chemical analyses, and today, using the electron microscope, even observed and photographed. Thus we know the molecular structures of RNA and DNA, of nylon, crystals, metals, and so on.

At the next level, molecules are composed of atoms which do not have the form and chemical properties of molecules, neither hydrogen nor oxgen by itself possessing the properties of water. Yet atoms have their own properties of atomic number and weight, accounting for their organization in the Periodic Table. Like molecules which are the basic units retaining the properties of substances, atoms are the smallest particles of an element that can exist either in isolation or in molecular combination, since they enter into and are the products of molecular interactions. Though the substances are different, the oxygen atoms are the same in H_2O, NaOH, and H_2SO_4.

Until the experiments of Rutherford, atoms were believed to be the indivisible, indestructible building blocks of the universe—if their existence was accepted at all. Then, as described in Chapter IV, by bombarding gold foil with alpha particles, Rutherford discovered that atoms are composite, consisting of a massive nucleus surrounded by orbiting electrons. At first the nucleus was thought to consist only of positively charged particles that Rutherford named protons, but in 1932 Chadwick confirmed the existence of a second nucleon, called the neutron, possessing a mass similar to the proton but having no electric charge. These discoveries enabled physicists to explain the atomic properties of number (consisting of the number of protons), weight (essentially the combined masses of the nucleons), charge (the normally neutral charge of atoms due to their equal number of negatively charged electrons and positively charged protons, with the neutrons being neutral in charge), and chemical properties (related to the number of electrons) that were previously inexplicable at the atomic level. Then Bohr explained the stability and characteristic spectral emission of atomic elements in terms of his theory of discrete electron orbits and quantum jumps. Again, for a time, it was believed that this conception represented the basic structure of the atom.

However, following the experimental methods of Rutherford, his student John Cockcroft, in collaboration with E. T. S. Walton, in 1932 succeeded in causing a nuclear transmutation by directing a beam of protons at the nucleus. At about the same time, the American Ernest O. Lawrence working with M. S. Livingston developed another type of particle accelerator, the cyclotron which, as in the experiments of Cockcroft and Walton, generated highly accelerated, high–energy protons that were used to penetrate, and thereby explore, the composition of the nucleus. Thus began the construction of a series of more powerful particle accelerators to probe deeper into the composition of subnuclear particles. As Pagels states:

> Physicists realized that with the cyclotron they now had an instrument, like Galileo's telescope, that could explore a new realm of nature—the subatomic microcosmos. Beyond the atom, beyond the nucleus, lay a virgin territory, a place never before seen. Physicist believed that in the nucleus lay the clue to the ultimate structure of matter and fundamental laws of nature.[18]

It was not until after the Second World War with the construction of a second generation of more powerful accelerators, the cosmotron at Brookhaven National Laboratory and the Bevatron at Berkeley, that the exploration of subnuclear particles could really begin. With beams of high–energy protons another, deeper level of matter was discovered consisting of the baryon group among the hadrons. Protons and neutrons, it turned out, were just two relatively stable hadrons out of innumerable unstable particles. But even the neutron decays into a proton, electron, and neutrino, so that the proton, electron, neutrino, and photon are considered stable (although at the present there is considerable research to determine if the proton decays). The liquid hydrogen bubble chamber was invented to facilitate the identification of the interactions of these elementary particles, while Richard Feynman devised diagrams to depict their impacts and disintegrations. In the sixties a new group of hadrons, the mesons, was discovered, along with "strange particles" carrying a new quantum number or charge, "strangeness." As expected, the hadrons exhibit novel properties and interactions, such as the integer spins of mesons and the fractional spins of baryons. In addition to these differences, the baryons are distinguishable from the mesons by the fact that the number of baryons produced by a collision is equal to the number that entered the collision, establishing the law of baryon conservation which does not apply to mesons. New charges, such as isotopic charge and strangeness, that are preserved in the interactions also were discovered to belong to the hadrons. It was these new conservation laws of charges that provided clues as to the structure of the hadrons.

For the unexpected but rapid proliferation of the hadrons aroused suspicion whether they, too, were composite. So in the early sixties, guided by the new conservation laws and the organization of the hadrons in a new kind of periodic chart (described in the previous chapter), Murray Gell–Mann detected mathematical symmetries among the hadrons that led to his formulation of "the eightfold way," an organizing of the hadrons that eliminated the chaos brought about by their proliferation. But then it was natural to ask why this eightfold classification was successful, because previously such classifications had always been indicative of a deeper and simpler organization of elements. The answer, proposed independently by George Zweig and Gell–Mann, consisted of a more basic level of fundamental particles out of which the hadrons might be constructed. Thus was born the theory of quarks with fractionál charges which, when conserved in the reactions of the hadrons, could account for the latter's conservation laws. Once again, a more powerful generation of particle accelerators was built in the late sixties and seventies, the Stanford Linear Accelerator (SLAC), The Fermi National Accelerator Laboratory (Fermilab), and the neutron interaction detector at CERN, along with accelerators in Italy and Germany, to confirm these theories.[19] As a consequence, though isolated quarks are not detectable, evidence for their location in hadrons has been inferred from the fact that various hadrons have been discovered based on the predictions of the quark theory. Such evidence has confirmed the existence of five quarks, with a sixth one expected. Again it is natural to ask, do these quarks constitute the basic, irreducible level of physical reality? Pagels' answer is as frank as possible:

In the future, additional levels may appear—perhaps governed by new laws of physics. We have no evidence that anything new must happen; it seems for all purposes that with the discovery of quarks we have reached the end of our journey. But there is an uneasy feeling among physicists that the trip is not over (*ibid.*, p. 207).

These developments indicate, in my opinion, that physical reality consists of a series of levels, each composed of distinct layers of entities with unique properties that account, to some extent, for the kinds of structures and interactions one finds on the succeeding higher levels. That they account only partially for the higher domains implies that new features emerge that are not fully explicable. The discovery of each of these levels depends upon modes of investigation correlated with increased intensities of energy. It is as if we were gazing into an enormous sphere, the interior structures of which changed as, aided by various instruments, we penetrated more deeply into its interior. The outer level is the most painterly, consisting of the richest diversity of qualities and forms, as well as the greatest variety of interrelations, while each successive level becomes less intricate and varied, though disclosing new substructures and interactions. This progressive decline in complexity is compensated for by an evident increase in simplicity, unity, and coherence among the elements, contributing to their greater intelligibility. Each of these contexts is somewhat autonomous, although discerned interactions and interconnections between levels provide the basis for deeper or fuller explanations. It is these necessary connections, discovered *a posteriori,* that constitute our scientific knowledge. Yet is seems to be a fundamental feature of this magnificent sphere that complete, continuous transitions from one level to another are neither perceptually nor conceptually feasible, new levels of entities emerging into prominence as we penetrate further with the aid of our instruments. Thus while deeper levels account in some ways for the higher levels, in the absence of a continuous transition we cannot fully explain how or why the higher contexts acquire their distinctive qualities, properties, and forms. This is the meta-physical picture of contextual realism, a conception that seems to be more consistent with the achievements of contemporary science than Kant's notion of a cognitive curtain separating phenomena from noumena, or the reduction of other levels to one basic level.

While this conception of inexhaustible, irreducible contexts or levels will probably be replaced in the future by a more adequate meta-physical model, in the meantime, rather than attempting to collapse reality into one dimension, which invariably blinds the investigator to the significance of other contexts, we should assume that the universe consists of an endless nexus of domains with innumerable structures and features. Experimentation and theory construction have been the most successful means so far in providing partial glimpses of this reality, as refracted through our cognitive–linguistic frames, but we should not assume these means are final. As Henry Stapp asserts:

> Such a view, though withholding the promise for eventual complete illumination regarding the ultimate essense of nature, does offer the prospect that human inquiry can continue indefinitely to yield important new truths. And these can be final in the sense that they grasp or illuminate some aspect of nature as it is revealed to human experience.[20]

Although we surely can expect further startling discoveries and fundamental breakthroughs in knowledge, there is little to indicate that we shall ever dispel the profound mysteries enshrouding our existence. This realization, while not fulfilling man's quest for absolute knowledge or ultimate truth, is nevertheless compatible with the notion of human inquiry as an unending adventure exploring the infinite recesses of nature.

> To ask and ask is but a child's way,
> To answer, that is man's fond play—
> But more than play I'll not contend,
> For question follows answer in the end.

Notes to Chapter IX

1. Cf. Henri Bergson, *Introduction to Metaphysics,* trans. by T. E. Hulme (New York: Library of Liberal Arts, [1903] 1913).
2. F. H. Bradley, *Appearance and Reality,* 2nd ed. (Oxford: Clarendon Press, 1897), p. 3.
3. I first used the term "contextualistic realism" in an article, "Science, Truth and Ordinary Language," in *Philosophy and Phenomenological Research,* September 1966, p. 36. The term is also the title of a more recent article in the same journal, June 1981, pp. 437–451. But just as the original term "pragmaticism" used by Peirce was shortened to "pragmatism," so the term "contextual" seems preferable to "contextualistic."
4. Cf. David Bohm, *Causality and Chance in Modern Physics* (Philadelphia: University of Pennsylvania Press, [1957] 1971), p. 137ff. I am particularly indebted to Bohm's description of the physical world as a matrix of "inexhaustible qualities" and infinite layers of interrelated though semi–autonomous "levels", described in the above book, for my own sketch of reality,' although I do not expect that the current "paradoxes" in quantum mechanics will be resolved, as were so many past scientific impasses, by uncovering a deeper level of subatomic phenomena (the "hidden variables" view), as Bohn argued in that book. Recently he has suggested that quantum theory is indicative of a new order of physical reality, an "implicate order." Cf. David Bohm, *Wholeness and the implicate order* (London: Routledge & Kegan Paul, 1980), chs. 5, 6.
5. Cf. Mario Bunge, *Metascientific Queries* (Springfield, IL: Charles C. Thomas, 1959), ch. V.
6. Cf. John Cowperwaithe Graves, *The Conceptual Foundations of Contemporary Relativity Theory* (Cambridge: MIT Press, 1971), pp. 17–18.
7. Karl R. Popper and John C. Eccles, *The Self and Its Brain* (New York: Springer International, 1977), pp. 554–555. Popper also refers to "hierarchical levels" in *Conjectures and Refutations: The Growth of Scientific Knowledge* (New York: Harper Torchbooks, 1968), p. 173.
8. Niels Bohr, *Atomic Theory and the Description of Nature* (Cambridge: Cambridge University Press, 1934), p.55.
9. Richard Schlegel, *Superposition and Interaction* (Chicago: University of Chicago Press, 1980), p. 269.

10. Karl R. Popper and John C. Eccles, *op. cit.,* p. 193.
11. David Bohm, *Causality and Chance in Modern Physics, op. cit.,* p. 100.
12. Cf. John Cowperwaithe Graves, *op. cit.,* p. 313.
13. An article in *The Washington Post,* "Physicist Freeman Dyson," April 9, 1984, B11.
14. Isidor Rabi, "Profiles—Physicist, II," *The New Yorker Magazine,* October 20, 1975. Quoted from Gary Zukav, *op. cit.,* p. 328.
15. Moritz Schlick, "The Turning Point in Philosophy," trans. by David Rynin, reprinted in A. J. Ayer (ed.), *Logical Positivism* (New York: Free Press, 1959), p. 54. The article, with its original German title, *"Die Wende der Philosophie,"* was first published in *Erkenntnis,* Vol. I, 1930-31.
16. For example, see Maurice Mandelbaum, *Philosophy, Science and Sense Perception* (Baltimore: Johns Hopkins Press, 1964), pp. 221ff, and Wilfrid Sellars, "Phenomenalism," in *Science, Perception and Reality* (New York: Humanities Press, 1963), pp. 96-97.
17. Cf. Paul M. Churchland, *Scientific Realism and the Plasticity of Mind* (Cambridge: Cambridge University Press, 1979).
18. Heinz R. Pagels, *The Cosmic Code* (New York: Simon & Schuster 1982), p. 197.
19. Cf. *ibid.,* pp. 193-225. I am indebted to Professor Pagels's excellent description of these experimental developments. All philosophers should read it!
20. Henry Peirce Stapp, "The Copenhagen Interpretation," *American Journal of Physics,* August, 1970, p. 1111.

Index

C

D

E

F

M

N

Y

Z